THE INVENTION OF DISCOVERY, 1500–1700

Literary and Scientific Cultures
of Early Modernity

Series editors:

Mary Thomas Crane, Department of English, Boston College, USA
Henry Turner, Department of English, Rutgers University, USA

For a decade now, *Literary and Scientific Cultures of Early Modernity* has provided a forum for groundbreaking work on the relations between literary and scientific discourses in Europe, during a period when both fields were in a crucial moment of historical formation. We welcome proposals that address the many overlaps between modes of imaginative writing typical of the sixteenth and seventeenth centuries—poetics, rhetoric, prose narrative, dramatic production, utopia—and the vocabularies, conceptual models, and intellectual methods of newly emergent 'scientific' fields such as medicine, astronomy, astrology, alchemy, psychology, mapping, mathematics, or natural history. In order to reflect the nature of intellectual inquiry during the period, the series is interdisciplinary in orientation and publishes monographs, edited collections, and selected critical editions of primary texts relevant to an understanding of the mutual implication of literary and scientific epistemologies.

The Invention of Discovery, 1500–1700

Edited by

JAMES DOUGAL FLEMING
Simon Fraser University, Canada

Routledge
Taylor & Francis Group

LONDON AND NEW YORK

First published 2011 by Ashgate Publishing

2 Park Square, Milton Park, Abingdon, Oxfordshire OX14 4RN
711 Third Avenue, New York, NY 10017

Routledge is an imprint of the Taylor & Francis Group, an informa business

First issued in paperback 2018

British Library Cataloguing in Publication Data
The invention of discovery, 1500–1700. --
 (Literary and scientific cultures of early modernity)
 1. Knowledge, Theory of--History. 2. Knowledge, Theory of,
in literature. 3. Discoveries in science--History--16th
century. 4. Discoveries in science--History--17th
century. 5. Scientific literature- History--16th century.
 6. Scientific literature--History--17th century.
 7. Literature and science--History--16th century.
 8. Literature and science--History--17th century.
 I. Series II. Fleming, James Dougal, 1968-
 809.9'336-dc22

Library of Congress Cataloging-in-Publication Data
Fleming, James Dougal, 1968-
 The invention of discovery, 1500–1700 / James Dougal Fleming.
   p. cm. -- (Literary and scientific cultures of early modernity)
 Includes bibliographical references and index.
 ISBN 978-0-7546-6841-1 (hardback)
1. Discoveries in science--History--16th century. 2. Discoveries in science--History--16th century. 3. Inventions--History--16th century. 4. Inventions--History--17th century. 5. Knowledge, Theory of--History--16th century. 6. Knowledge, Theory of--History--17th century. 7. Renaissance. I. Title.
 Q180.55.D57F57 2011
 509--dc22

                                                                                              2010052524

ISBN 978-0-7546-6841-1 (hbk)
ISBN 978-1-138-38277-0 (pbk)

Contents

Notes on Contributors

Michael Booth teaches Shakespeare at Northeastern University. His primary research interest is in cognitive approaches to the study of poetry (and other forms of literary and cultural innovation), particularly the theory of conceptual blending.

Piers Brown is a SSHRC post-doctoral fellow at the University of York. He recently completed a PhD on Donne and early modern astronomy at the University of Toronto. His chapter in this volume was completed during a Francis Bacon fellowship at the Huntington Library.

Travis DeCook is Assistant Professor of English at Carleton University. His current research focuses on the culture of eschatological politics in early-modern England.

Louise Denmead teaches early-modern literature in the Department of English, University College, Cork. Her research interests are in Renaissance drama, representations of race on the early-modern stage, and postcolonial and feminist theories. She is currently working on a book entitled *The Myth of Ixion: Imagining Race and Interracial Sex in the Renaissance*.

James Dougal Fleming is Associate Professor of English at Simon Fraser University. His interests, and publications, are in the intersections of early-modern textuality and modern hermeneutics.

Vincent Masse is Assistant Professor in the French Department of Dalhousie University. His PhD thesis (University of Toronto, 2009) focused on the insertion of "New Worlds" in early printed French discourses. In 2009–10 he worked on a SSHRC post-doctoral research project at the Université Paris-IV Sorbonne, on Guillaume Postel (1510–81), the invention of orientalism, and the esoteric and messianic foundations of modern colonialism.

Steven Matthews is Associate Professor of History at the University of Minnesota, Duluth. His specialization is in the interaction between theology and science in the early-modern period.

Ryan Netzley is Assistant Professor of English at Southern Illinois University, Carbondale. He is the editor of *Acts of Reading: Interpretation, Reading Practices, and the Idea of the Book in John Foxe's* Actes and Monuments (University of

Delaware Press, 2010), and the author of *Reading, Desire, and the Eucharist in Early Modern Religious Poetry* (University of Toronto Press, 2011).

Pietro Daniel Omodeo obtained a PhD in Philosophy in 2008 at the University of Turin (Italy). He was recipient of the 2009 Kristeller-Popkin Travel Fellowship, sponsored by the *Journal of the History of Philosophy*. In 2010, he was a scholarship holder of the Herzog August Library in Wolfenbüttel (Germany) and of the Fritz Thyssen foundation at the Research Library of Erfurt-Gotha (Germany). He has also been elected for a 2010–11 resident fellowship at the Max Planck Institute for the History of Science in Berlin, with a project on the history of astronomy. His main field of research is the philosophical and scientific culture of the Renaissance with particular focus on Cusanus, Copernicus and Bruno.

Anthony Russell is Associate Professor of English and Comparative Literature at the University of Richmond. He has published on John Donne, Rabelais and Folengo, and Dante's *Vita Nuova*. He is currently at work on a book entitled *"The Conspiracy of Our Spatious Song": Magic, Vitality, and the "Spiritus Phantasticus" in Late Medieval and Renaissance Poetry*.

Jacqueline Wernimont is Assistant Professor of English at Scripps College in Claremont, California. She is at work on a monograph exploring representations of possibility in early-modern poetry and mathematics, and the import of those representational modes for our theories of mimesis. She teaches early-modern literature and history of science, as well as digital humanities.

Acknowledgements

This book grew out of a panel I organized under the rubric "Theories of Discovery" at the Renaissance Society of America (RSA) conference in Chicago, 2008. Thanks to Jacqueline Wernimont and Michael Booth for joining me on the panel; to Jonathan Sawday for acting as respondent; to Mary Baine Campbell for chairing; and to all who attended and asked questions in what was a very lively session. It is regrettable that RSA does not make more room for this sort of thing. Work on the book devoured much of—or rather, was made possible by—my study leave from Simon Fraser University; for which (the leave) I am grateful. (Note to wife and children: at least it started and ended at the beach.) Some of the research on which my own work here is based was supported by the Social Sciences and Humanities Research Council of Canada (SSHRC). Special thanks to Steve Matthews, who told me this book would edit itself.

for Troy

the new

Introduction
The Invention of Discovery, 1500–1700

James Dougal Fleming

The early-modern period between Columbus and Newton used to be called "the Age of Discovery." This was the era when European Man (invariably, Man) found the world anew. Or rather, he found new worlds—two of them. One was geopolitical: the Indies, east and west. The other was epistemological: modern natural science. The significance of these worlds was precisely that they were *found*, in the objective nature of their truth, rather than made. Discovery meant liberation from superstition and a positive encounter with decisive facts. The result was Jefferson, Lavoisier, Kant—in a word, Enlightenment.[1]

That story has few tellers anymore. *Invention*, as opposed to discovery, is now the academic leitmotif of the early-modern.[2] That rebranding, moreover, carries a critical sense. Far from merely finding new aspects of the world as it was—it is now argued—the expansionist Europe of the sixteenth and seventeenth centuries *made* its new worlds, in many aspects, as these have come to be. This is not only a trite observation about modernity and innovation. It is historicized skepticism about innovation in modernity. For the story of wondrous finding, as a model for what basically happened during the early-modern period, can be treated as both analytically suspect and ethically turpid. The resulting turn from

---

[1] See Daniel Boorstin, *The Discoverers* (New York, 1983); Wilcomb E. Washburn, *The Age of Discovery* (Washington, DC, 1966); Paul Hermann, *Great Age of Discovery* (New York, 1958); and Arthur Percival Newton, *The Great Age of Discovery* (London, 1932).

[2] See Lorna Hutson, *The Invention of Suspicion: Law and Mimesis in Shakespeare and Renaissance Drama* (Oxford and New York, 2007); Sean Keilen, *Vulgar Eloquence: On the Renaissance Invention of English Literature* (New Haven, 2006); Andrea Frisch, *The Invention of the Eyewitness: Witnessing and Testimony in Early-Modern France* (Chapel Hill, 2004); John E. Crowley, *The Invention of Comfort: Sensibilities and Design in Early-Modern Britain and Early America* (Baltimore, 2001); Evelyn Lincoln, *The Invention of the Italian Renaissance Printmaker* (New Haven, 2000); Judith Veronica Field, *The Invention of Infinity: Mathematics and Art in the Renaissance* (Oxford and New York, 1997); Lynn Hunt (ed.), *The Invention of Pornography: Obscenity and the Origins of Modernity, 1500–1800* (New York, 1996); Pamela Benson, *The Invention of the Renaissance Woman: The Challenge of Female Independence in the Literature and Thought of Italy and England* (University Park, 1992); Murray Pittock, *The Invention of Scotland: The Stuart Myth and the Scottish Identity, 1638 to the Present* (London, 1991); etc.

discovery to invention, as a ruling trope for early-modern studies, owes much to phenomenological critique, even more to a Foucauldian notion of social construction; and is consistent with a broad current of postmodernist revisionism. The discovery of America, the anatomization of sexuality, and the empiricization of natural philosophy are just a few of the traditionally early-modern (alleged) objectivities that have been reconsidered in this way.[3]

No doubt, for early-modern studies, this is a long-term methodological advance. Yet it brings with it several dangers. For one thing, the analytic turn from discovery to invention—from objective finding, to cultural making—tends to be presented as an interesting *peripeteia* (or reversal of fortune). In other words, it is supposed to *matter*, more-or-less by definition, if discovery turns out to be invention: if a given early-modern finding, handed down as such since the Enlightenment, can really, and more properly, be understood as a projection or construction. This mattering, however, runs the risk of reifying the starting-position that the analysis was supposed to reduce: namely, the traditional hierarchization of discovery over invention. If the significance of early-modern invention is primarily that it results from deconstructing discovery, then it is discovery, and not invention, that grounds a claim on our interest. So it is with any peripeteia. Reversal of fortune depends upon, and re-projects as normative, that fortune. Ending the age of discovery, perhaps, is not quite as easy as it seems.

For another thing, early-modern European culture, as we know, did not even assume a stable distinction between invention and discovery, making and finding. This point is usually presented as a classicizing rhetorical one, on a semantic basis: Latin *invenire* means "to find," and in Renaissance rhetoric the first step is "invention," which means finding something to say—often, and indeed preferably, in a pre-existing and model text. In this respect, the old "age of discovery" model imposed an anachronistic discretion, from an Enlightenment perspective, on a highly synthetic and untidy early-modern category. Yet the new "age of invention" model runs the risk of re-imposing that discretion, just from the other way around. To place invention over discovery, as much as to place discovery over invention, is still to be dealing with opposing terms. But it is not clear that the early-modern period dealt with them in that way.

At the same time—and this, perhaps, we do not quite know—the Enlightenment got its idea of early-modern discovery from at least some aspects and motifs of the preceding period. It is possible to observe a diachronic process via which various departments of Renaissance culture, following motivations of their own, pried the idea of invention apart from the idea of discovery, and subordinated the former to the latter—historically, epistemologically, and hermeneutically.[4] This became

---

3   See Edmundo O'Gorman, *The Invention of America* (Bloomington, 1961); Thomas Laqueur, *Making Sex: Body and Gender from the Greeks to Freud* (Cambridge, MA, 1990); and Steven Shapin and Simon Schaffer, *Leviathan and the Air Pump* (Princeton, 1985).

4   Recent studies consistent with an observation of this process include Mario Biagioli, *Galileo's Instruments of Credit: Telescopes, Images, Secrecy* (Chicago, 2006);

the binary that the Enlightenment inherited from the Renaissance, and handed down, in turn, as a triumphant legacy of modernity. For us postmoderns, thinking outside that legacy is likely to be difficult. Yet thinking about it, by that very token, becomes an urgent task. It may be that discovery, precisely *as* a period invention, remains the category we need to focus on, if we are to understand early-modern geopolitical and epistemological developments. The *invention of discovery*, in sum, may be the aboriginal version of all the "inventions" that have filled postmodern scholarship on the early-modern.

From Translation to Empire

Consider humanism. The *translatio* of classical letters, beginning in the fourteenth century, but extending well into the seventeenth, entailed the recovery of classical texts. This entailed, in turn, journeys to and excavations within far-flung and neglected libraries. Yet classical materials had not languished in these monastic collections just because people had not managed to find them. They had languished, rather, because people had, for the most part, not bothered to look. The potential discovery of classical originals had simply not occurred, at least not in a systematic or paradigmatic way, to a medieval culture that was frequently satisfied with the colorful misprisions of its own redactions.[5]

More to the point, humanism itself motivated discovery by *translatio*; it did not motivate *translatio* by discovery. Petrarch, one of the pioneers of the philological safari, nonetheless placed its signature activity very low on the intellectual totem pole. "The man who finds a gem inside a fish," he writes, "is not a better, but a more fortunate, fisherman." The discovery of antiquities by vinediggers and foundation-builders, "happily bedazzled by the unexpected brilliance of hidden gold," is no cause for wonder or admiration. Rather, it is the learned expert, recognizing the coin or text in its ancient and ongoing meaning, who is laudable.[6] For Petrarch, discovery is contemptible insofar as it is accidental—in Aristotelian

---

Ingrid D. Rowland, *The Scarith of Scornello: A Tale of Renaissance Forgery* (Chicago, 2004); Leonard Barkan, *Unearthing the Past: Archaeology and Aesthetics in the Making of Renaissance Culture* (New Haven, 1999); and James Bono, *The Word of God and the Languages of Man: Interpreting Nature in Early Modern Science and Medicine, Volume 1: Ficino to Descartes* (Madison, 1995).

[5]   See Harry Vredeveld, "Deaf as Ulysses to the Siren's Song: The Story of a Forgotten Topos," *Renaissance Quarterly*, 54/3 (2001): pp. 846–82; C.A. Patrides, *Premises and Motifs in Renaissance Thought and Literature* (Princeton, 1982), pp. 105–23 and 152–81; Jean Seznec, *The Survival of the Pagan Gods: The Mythological Tradition and Its Place in Renaissance Humanism and Art*, trans. Barbara F. Sessions (Princeton, 1972); and D.C. Allen, *Mysteriously Meant: The Rediscovery of Pagan Symbolism and Allegorical Interpretation in the Renaissance* (Baltimore, 1970), pp. 215–23.

[6]   Quoted in Barkan, *Unearthing*, p. 1; and see pp. 28–9.

terms, non-substantial, not part of the way things are. Indeed, Petrarch's image of dumb digging luck directly recalls one of Aristotle's for ontological accidence (as opposed to substance): finding treasure while "digging a hole for a plant."[7] Of course, Petrarch's own famous discoveries, notably of Cicero's letters, did much to establish the humanist trope of philology *ad fontes* (toward the sources).[8] Nonetheless, his foundational insistence on the intellectual primacy of learned *recognition*—judgment and insight, based on foreknowledge—is an indication that the canonization of the discovery-trope was something that humanism itself had to work at.

Protestantism, in the sixteenth century, brought the secular trope into the sacred realm. Here, philology *ad fontes* came to mean (1) recovery of the Bible from beneath the accretions of ecclesiastical authority and patristic representation; and (2) recovery of Biblical intensions, in their original languages, from beneath the interpretations of canonical Latin (i.e., the Vulgate Bible). The latter move was then supposed to support the former, as the intensions of scripture could be universally vernacularized through new and, in principle, hermeneutically transparent translations.[9] Against the Protestant project, and despite a sympathetic hearing from such figures as Erasmus, stood the Roman Catholic idea of *tradition*: the "dogmatic unity" of authoritative exegesis, handed down under the validating aegis of the Holy Spirit.[10] At the Council of Trent, the Roman Church declared that the Vulgate Bible was to be considered authentically correct, and, indeed, incorrigible. Why? Because it had been "for so many ages allow'd of."[11] The very thing that prompts Protestants to attempt the discovery of the unredacted or uninterpreted Bible—the historical fact of its attenuation through generational transmission—prompts the Tridentine Church to hold on to its established chain of redactions and interpretations. This is not just a base or wicked refusal of a common-sense procedure—the procedure of discovery. Rather, the Tridentine position is a principled rejection of discovery, based on an alternative philology. For Rome, it

---

[7]   Aristotle, *Metaphysics*, in *The Basic Works*, ed. and intro. Richard McKeon (New York, 1941), p. 777.

[8]   See Carol E. Quillen, *Rereading the Renaissance: Petrarch, Augustine, and the Language of Humanism* (Michigan, 1998), and William J. Kennedy, *Authorizing Petrarch* (Ithaca, 1994).

[9]   Useful recent accounts are in Kevin Killeen and Peter Forshaw (eds), *The Word and the World: Biblical Exegesis and Early-Modern Science* (New York, 2007); Ariel Hessayon and Nicholas Keene (eds), *Scripture and Scholarship in Early-Modern England* (Aldershot, 2006); David Bagchi and David C. Steinmetz (eds), *The Cambridge Companion to Reformation Theology* (Cambridge and New York, 2004); and Richard A. Muller and John L. Thompson (eds), *Biblical Interpretation in the Era of the Reformation* (Grand Rapids, 1996).

[10]   Hans-Georg Gadamer, *Truth and Method*, 2nd rev. ed., trans. rev. Joel Weinsheimer and Donald G. Marshall (New York, 2004), p. 176.

[11]   *Canons and Decrees of the Council of Trent* (London, 1687), p. 13.

is basically a *good sign* if a given scriptural reading, whether or not supported by *ad fontes* philology, has been handed down, without God's prevention, throughout Christian history. Protestantism, a controversial (to say the least) innovation of the period, was controversial in part because of its countervailing commitment to the discovery-trope.

At the same time, as with humanism, we find a tension or confusion on this point within Protestantism itself. Philologically, the Reformers (Tyndale, Luther, Calvin) are zealous discoverers: they seek, and will only stop at, the authentic word of God. By that very token, however, that is where they wish to stop.[12] Hermeneutically, the Reformers do not wish to prosecute discovery. Their watchword is the literal or "historical" meaning of scripture; and they excoriate the medieval Catholic allegorists, who claim to be able to divine (as it were) God's occluded intensions.[13] To be sure, Protestant literalism admits two traditional classes of exceptions: typology (New Testament revelation as the real meaning of Old Testament history); and accommodation (scriptural representations as what human beings can understand, rather than what really is).[14] These exceptions, however, have strong theological and/or hermeneutic justifications. Absent any good reason to depart from the literal or self-evident sense of scripture—if it is neither incoherent on its face, nor impeding the glory of God—Protestant authorities are loath to do so.

Luther, for example, in his exegesis of Genesis 3:14 (God's punishment of the serpent for his role in the Fall), asserts that there is no valid reason to read "dust shalt thou eat" allegorically or figuratively. Literally, the line records the Almighty's curse on the luckless genus of snakes; and such a divine utterance, surely, must be totally

---

[12]   For the difficulty of stabilizing the stopping-place, however, see Scot Mandelbrote, "English Scholarship and the Greek Text of the Old Testament, 1620–1720: The Impact of Codex Alexandrinus," and Nicholas Keene, "A Two-Edged Sword: Biblical Criticism and the New Testament Canon in Early Modern England," in Hessayon and Keene (eds), *Scripture and Scholarship*, pp. 74–93 and 94–115.

[13]   See Sarah Hutton, "Iconisms, Enthusiasms and Origen: Henry More Reads the Bible," in Hessayon and Keene (eds), *Scripture and Scholarship*, pp. 192–207; Peter Harrison, "Reinterpreting Nature in Early Modern Europe: Natural Philosophy, Biblical Exegesis and the Contemplative Life," in Killeen and Forshaw (eds), *The Word and the World*, pp. 25–44; Donald McKim (ed.), *The Cambridge Companion to John Calvin* (Cambridge, 2004) and *The Cambridge Companion to Martin Luther* (Cambridge, 2003); D.C. Allen, *Mysteriously Meant*, pp. 239–42; and Edgar Wind, "The Revival of Origen," in Dorothy Miner (ed.), *Studies in Art and Literature for Belle da Costa Greene* (Princeton, 1954), pp. 412–24.

[14]   See Peter Harrison, *The Bible, Protestantism and the Rise of Natural Science* (Cambridge, 1998); Robert S. Westman, "The Copernicans and the Churches," in David S. Lindberg and Ronald Numbers (eds), *God and Nature: Historical Essays on the Encounter Between Christianity and Science* (Berkeley, 1986), pp. 76–113, esp. 90–91; and Arnold Williams, *The Common Expositor: An Account of the Commentaries on Genesis, 1527–1633* (Chapel Hill, 1948).

and permanently effective, in exactly its own terms. Therefore, Luther concludes, *as a matter of zoological fact*, that snakes eat dirt.[15] It has often been suggested that Protestant literalism, in one way or another, facilitated the emergence of modern natural science.[16] Luther's account of serpentine diet—and one could give many other examples—suggests an alternative view. In any case, the current point is that the discovered text of God's word does not necessarily serve the Reformation mind as a site for further, interpretative, discovery. Protestantism does not provide a model of hermeneutics based on the signature trope of humanist philology.

That move is left to the third innovation we have to consider in the period's intellectual culture: (early) modern natural science. Literalism (admitting the established exceptions), the hermeneutic accompaniment of Protestant philology, finds its scientific analog only in observation-reports: saying what was seen. The characteristic and transformative claim of the new science, however, is precisely that what was *seen* is no more than an index to what is *true*. Metals, for example, may seem like *sui generis* substances; but laboratory practice shows that they can best be explained as secondary functions of largely unobservable, and genuinely substantial, corpuscles. Similarly, a drop of water, though evidently lifeless, can be made to reveal a teeming micro-world. Even more startlingly—and far more radically than the hypothetical epicycles of Ptolemaic astronomy—geocentric impressions can be reduced to heliocentric facts.

The last two examples, admittedly, involve technological innovations (the microscope and telescope) that consist in nothing other than an augmentation of observation. That is to say, however, that the new-scientific trope of discovery, unlike its humanist and Protestant antecedents, *does not stop* at the clear determination of its data-set. Rather, the same process that brought the investigator to the data—the process of discovery—must be repeated upon and within the data, in order to find out its true meaning. The book of nature, to adopt the ubiquitous period metaphor, must first of all be discovered; but then it must submit to further discovery of its intelligible structure—its sense, in semantic terms. The meaning of evidence, as a *standing rule or expectation of scientific practice*, must be recovered from within, and even against, its initial presentation.

In a word: *allegory*—not literalism—typifies the new-scientific conception of data-interpretation. The way things seem, even on close observation, is no better than an occluded guide to the way they truly are. Of course, the way things truly are, insofar as it can be grasped or known, must then be *protected* from further interpretative penetration, lest scientific factuality produce infinite regress. That is to say, however, that scientific hermeneutics leads to a conception of factuality that

---

[15]   Martin Luther, *Lectures on Genesis. Chapters 1–5*, in *Luther's Works*, ed. Jaroslav Pelikan (Saint Louis, 1958), vol. 1, p. 187.

[16]   See Stephen Gaukroger, *The Emergence of a Scientific Culture: Science and the Shaping of Modernity, 1210–1685* (Oxford, 2006), pp. 135–48; Harrison, *The Bible, Protestantism and the Rise of Natural Science*; and I.B. Cohen (ed.), *Puritanism and the Rise of Modern Science: The Merton Thesis* (New Brunswick, 1990).

must exempt itself from its own generative procedures. It is striking that two giants of the new science, Galileo and Newton, when they involve themselves in what we would now consider a non-scientific area of interpretative activity—Biblical scholarship—insist (1) on highly allegorical or at least non-literal readings of crucial passages or issues (scriptural geocentrism, for Galileo; Trinitarianism, for Newton), even as they insist (2) on the absolute and literal factuality of the very non-evident, even paranoid, readings they thereby achieve.[17] "This far—but no further" is the rule of their hermeneutics. Both its productivity, and its instability, are derived from a theoretical and methodological alignment, if not isomorphism, between valid data-interpretation and the penetration of evidentiary appearances.

We can call this general conception an *hermeneutics of discovery*.[18] It cannot be assumed for pre-scientific natural philosophy. After all, Renaissance scholasticism tended to place the recondite core of a natural phenomenon—its substantial form, or essential being—*beyond* empirical discovery.[19] True, the period's Neoplatonism opened up phenomenal essences to the possibility of natural-philosophical knowledge. It did so, however, precisely by conceiving of essences as engines of their own spontaneous manifestation.[20] The same idea informs Neoplatonic psychology—which incorporates, by that very token, physiology and physiognomy (since the soul determines, and can therefore be read off of, the body); and even astronomy. The Ficinian cosmologist Patrizi, as Kepler complained, rejected both chief world systems (the Copernican and the Ptolemaic), not because he had a better one, but because he rejected the necessity of any. For Patrizi, the erratic and supposedly puzzling planets "move amongst the fixed stars ... exactly as they appear to."[21] Natural self-evidence produces an imperative to what we can call an *hermeneutics of recognition*: not the penetration of appearances, but their contemplation and integration.[22] Perhaps this is somewhat like Petrarch's emphasis

---

[17]　For Galileo, see J.D. Fleming, "Making Sense of Science and the Literal," in Killeen and Forshaw (eds), *The Word and the World*, pp. 45–60; Ernan McMullin, "Galileo on Science and Scripture," in Peter Machamer (ed.), *The Cambridge Companion to Galileo* (Cambridge, 1998), pp. 271–347; and William Shea, "Galileo and the Church," in Lindberg and Numbers (eds), *God and Nature*, pp. 114–35. For Newton, see Rob Iliffe, "Friendly Criticism: Richard Simon, John Locke, Isaac Newton and the Johannine Comma," in Hessayon and Keene (eds), *Scripture and Scholarship*, pp. 137–57; Maurizio Mamiani, "To Twist the Meaning: Newton's *Regulae Philosophandi* Revisited," in Jed Z. Buchwald and I.B. Cohen (eds), *Isaac Newton's Natural Philosophy* (Cambridge, MA, 2001), pp. 3–14; and J.E. Force and R. Popkin (eds), *Newton and Religion: Context, Nature and Influence* (Dordrecht and Boston, 1999).

[18]　See J.D. Fleming, *Milton's Secrecy and Philosophical Hermeneutics* (Aldershot, 2008), pp. 1–4.

[19]　See Chapter 4, below.

[20]　See David Freedberg, *The Eye of the Lynx: Galileo, His Friends, and the Beginnings of Modern Natural History* (Chicago, 2002).

[21]　Gaukroger, *The Emergence*, p. 100.

[22]　See Fleming, *Milton's Secrecy*, pp. 25–8 and 123–9.

on learned appreciation. It is very unlike Galileo's, or Boyle's, or Leeuwenhoek's emphasis on empirical penetration.

And what, in the end, is the argument for such penetration? It is power and convenience. Practical and political motivations, as Bacon himself explicitly recognized, underwrote the allegedly transcendent standards of emergent science. This point has been reinforced, in recent years, from the somewhat surprising quarter of scholarship on the Iberian empires. Anglocentric narratives of the scientific revolution have tended to marginalize Spanish and Portuguese contributions, limiting them to a retrograde and intellectually tawdry militarism. It is now apparent, however, that the project of imperial conquest prompted Iberian adventurers to anticipate British technological innovations, in some cases, by centuries.[23] Perhaps this is not really very surprising after all, coming from a culture that almost self-consciously invented discovery, as the category through which to make sense of the bizarre cosmographical accident that came to be called the New World.[24] But the point is that an hermeneutics of discovery did not precede the conquests, as the established method for obtaining the kind of knowledge that is called power. Rather, it emanated from, and was validated by, the conquests, as the method by which power had been obtained, and therefore was called knowledge.

Obviously, the above sketch, necessarily both selective and generalizing, cannot establish that the early-modern period invented the hermeneutics of discovery. It can, however, open up that possibility as a question or hypothesis. Under the latter, early-modern science emerges as a crucial topic. Yet precisely because of the intellectual hegemony of *modern* natural science, critical examination of its hermeneutic origins entails wider intellectual and cultural intersections.

Inventions and Discoveries

This collection begins, accordingly, with a series of chapters that juxtapose the new-scientific trope of discovery with some of its extra-scientific matrices. Piers Brown leads off, in Chapter 1, by examining the role of travel narrative in both Johannes Kepler's and Galileo Galilei's presentations of their astronomical discoveries. Brown argues that a journey-trope does more to construct the natural-philosophical objectivity of the new astronomy than does the more familiar trope of the scientific hunt. For hunting or searching is teleological, to the extent that one has to know what one is looking for. But journeying can be *accidental*: one bumps up against the facts. To describe one's findings in terms of travel, accordingly, is

---

[23]   See Antonio Barrera-Osorio, *Experiencing Nature: The Spanish American Empire and the Early Scientific Revolution* (Austin, 2006); Jorge Cañizares-Esguerra, *Nature, Empire, and Nation: Explorations of the History of Science in the Iberian World* (Stanford, 2007); and Daniela Bleichmar, Paula de Vos, Kristin Huffine, and Kevin Sheehan (eds), *Science in the Spanish and Portuguese Empires, 1500–1800* (Stanford, 2009).

[24]   See O'Gorman, *The Invention*, pp. 9–25.

precisely to construct them as very strongly factual. Yet this construction depends on an almost ludic randomness that is in tension with the strict objectivity that it is supposed to support. Brown has illuminated a paradox in the relationship between the new-scientific astronomy and its data: on the travel-trope, it is precisely because the truth of the cosmos *could have been discovered differently* that the way it actually was discovered is the only way it can possibly be.

Steven Matthews, in the following chapter, asks how and why new-scientific discovery itself came to be discovered, at least insofar as Sir Francis Bacon was concerned. Matthews's answer is that Bacon discovered discovery in theology: specifically, the negative theology of Pseudo-Dionysius. This gave Bacon the original of his method of negative instances, canonized today as "eliminative induction." Although scholars have been dubious about the natural-philosophical efficacy of Bacon's method—whether or not it actually helps the proto-scientist to understand any phenomena—Matthews argues that its real role is to ground natural-philosophical efficacy as such. Baconian science emerges by appropriating and displacing Pseudo-Dionysian contemplation of the celestial hierarchies. Matthews's argument is a startling contribution to the recent trend of reading early-modern science and religion as coeval, rather than opposed. The effect is not to deconstruct Baconian discovery, subversively or reductively, but to show what it is in its own terms: a methodological innovation, directed toward science in a very broad sense, and relying on highly abstract, even transcendent, postulates.

Michael Booth, in this book's third chapter, takes us back down to brass tacks—or grey matter. Booth approaches the work of the Elizabethan astronomer, mathematician, and comparative linguist Thomas Harriot through "blend-theory," a recent innovation of cognitive science. The implication of his analysis is that a binary of discovery and invention may be far too coarse a "blend" for the kinds of intellectual and period complexities that Harriot embodies. A man who deconstructed his own literacy in order to construct Algonquian grammar; whose mathematical innovations remained misunderstood until the twentieth century; and who made, but failed to claim, some of the period's most significant astronomical discoveries is a man who challenges multiple modern and postmodern assumptions about the departments and hierarchies of knowledge. To meet Harriot's challenge, per Booth's argument, is not to go down the ever-dwindling paths of postmodernist constructivism. It is, rather, to recognize that neither discovery nor invention can fully map the productive terrain on which data turn to knowledge.

My own contribution, appearing here as Chapter 4, varies from the emergence of the new science to the disappearance of the old. The doctrine of occult qualities, in neo-Aristotelian *scientia*, is a familiar topic to students of this area. I argue, however, (1) that the doctrine has been conflated with period esotericisms from which it is actually distinct; and (2) that its theoretical implications, as a result, have not properly been grasped. The occult-qualities doctrine draws a sharp epistemological line between limited and unlimited conceptions of natural-philosophical discovery. *Scientia*, I argue, is on the one side of that line; early-modern esotericism (whether Neoplatonic, Paracelsian, or alchemical) is on the other. Insofar as an unlimited writ of discovery suggests an unlimited scope for human knowledge, it propagates, arguably, epistemological incontinence. My chapter, in sum, is an attempt to make neo-Aristotelian thinking about occult qualities look good. For the latter avoids the identity of the unknown and the knowable that is such a destabilizing feature of early-modern esotericism—and, perhaps, of modern natural science.

Of course, whenever one talks about what is science, and what isn't, one runs up against countervailing modern conceptions of creativity or art. Anthony Russell, however, in Chapter 5, reminds us of the continuity between poetics and proto-science in early-modern Neoplatonism. Examining the concept of *ingegno* or genius in Ficino, Persio, and Campanella, Russell finds that it is (supposed to be) an innate and visceral capacity of the spirit or *pneuma*. This allows the esoteric magus to perceive the cohesive dynamism of the created world. As such, *ingegno* is exactly what John Donne laments and craves in his gloomy visions of an incoherent world—shattered, it seems, by the premature death of an obscure young woman whom Donne never even met (Elizabeth Drury). Yet the bathos of Donne's over-regretting, in the *Anniversaries*, perhaps suggests that the poet is drawing attention to natural magic as precisely what the early-modern world needs now. The poet and the proto-scientist, Russell's analysis suggests, join hands in the early-modern conception of esoteric genius.

The period nexus of creative insight and empirical work also occupies Pietro Omodeo, in Chapter 6. Omodeo takes a new look at the Copernican scene of the early seventeenth century. He finds it to be eclectic—composed of diverse and even conflicting "Copernicanisms"—and far from definitively committed to an epistemological rule of evidence. Picking up on Kepler's well-known preference for aprioristic science, Omodeo turns to the relatively neglected Savoy astronomer Giovanni Battista Benedetti. For Benedetti, large-scale cosmological questions, such as the potential infinity of space, were highly tractable of rational, as opposed to empirical, solution. Indeed, he seems to have believed that "rational possibility," in such a case, puts the onus on *dis*proof, rather than on proof. This quasi-idealist view, moreover, goes hand-in-hand with Benedetti's position—shared, needless to say, with Galileo and other prominent Copernicans—that astronomers and mathematicians really counted as natural philosophers.

Jacqueline Wernimont, in Chapter 7, turns to the period's great genius of rational possibility: René Descartes. Wernimont considers Descartes's early text *The World* as an exercise in what is now called possible-worlds theory. She argues, however, that the value of the exercise was precisely that it provided a template for empirical reality. Descartes's reader is to "discover" the validity of natural-philosophical mechanism in *The World*; s/he is then to discover the validity of natural-philosophical mechanism in the world. Strikingly, Descartes took the category of *fiction* about possible world-systems seriously enough to suppress his text from publication—at just about the time that Galileo was publishing a text, which he probably should have suppressed, because it was only a fiction (a dialogue) about possible world-systems.[25] Wernimont's goal, however, is not to read *The World* as an example of Descartes's cultural and political astuteness. It is, rather, to read *The World* as an example of "what 'discovery' entailed for the early-modern." For Descartes, as for Benedetti, empirical discovery entailed the mental production of "rational possibility." It entailed, in a word, invention.

With that, we leave any simplistic or reductive view of the period's empiricism behind. In this collection, accordingly, we leave behind the category of science, in the modern sense. Yet the tangle of discovery and invention leads to other areas of early-modern intellectual culture. Ryan Netzley, in Chapter 8, takes up the role of number in the Protestant martyrology, and apocalyptic eschatology, of John Foxe. Netzley finds Foxe, in his compendious *Actes and Monuments*, doing what one can only call creative numerology in order to make ecclesiastical history fit the Book of Revelation. The numbers Foxe comes up with are transparently invented. Yet the martyrologist seems to consider them significant—exegetic discoveries— precisely on the basis that they *were* invented. Foxe's attitude is all the stranger, as Netzley points out, given that he completely ignores the esoteric or cabbalistic meaning that his invented/discovered numbers, in some cases, have. Far from finding the arithmetic of Providence concealed in scripture, Foxe prefers to find it manifest, and available for recognition.

Travis DeCook, in Chapter 9, similarly opposes the modern hermeneutics of discovery to an early-modern, and English, and Protestant, hermeneutics of recognition. DeCook finds the seventeenth-century antiquarian and *via media* Protestant Thomas Fuller wrestling with discovery as productive of historiographic and theological error. On the one hand, Fuller criticizes radical Protestants for thinking they can—or need to—discover a pure or utopian state of the church. On the other hand, Fuller criticizes his own antiquarian profession for propagating an idolatrous fascination with discovery as such. His goal is to prevent belief, as DeCook nicely puts it, in "the pastness of the past." But what is the alternative? As Gadamer might say, it is genuinely hermeneutic: not mere discovery of the things that were, but engaged understanding of how they address the things that are—including the understanders.[26]

A number of the contributors to this book use literary categories to examine early-modern discovery. In Chapter 10, Louise Denmead examines discovery in early-modern (English) literature. Specifically, Denmead examines the popular dramatic device of the "bed-trick," in which a prospective sexual partner is swapped with an impostor, the subterfuge being subsequently revealed. Using as her main proof-text John Fletcher's *The Knight of Malta* (1617), Denmead asks what happens to this comic anagnorisis when the sexual impostor is herself (or himself) occluded by race, class, and gender. The myth of Ixion, in which a would-be lover of Juno clasps only a cloud in her form, allows Denmead to unpack the

---

[25]   See Maurice A. Finocchiaro (trans. and ed.), *Galileo on the World Systems* (Berkeley, 1997).

[26]   See Gadamer, *Truth and Method*, pp. 267–304.

full paradoxicality of bed-tricks involving black and (in almost all her examples) female servants. These involve both the dupe of the bed-trick, and the play's putative audience, in an almost uncharacterizable mixture of sudden recognition, and willful ignoring.

Finally—or perhaps one should say, conclusively—we move from texts to books. Vincent Masse, in Chapter 11, treats the motifs of discovery and newness in the print culture of early-modern France. Masse finds these motifs in increasing proliferation as the period goes on; he also finds that "claims of revolutionary content," in the French book trade, are far from reliable. From Masse's impeccably-researched argument we can infer that the discovery of novelty, over the course of our period, developed as a lucrative marketing strategy for publishers and booksellers. We can also infer, however, that the book trade itself actively and willfully participated in the construction of these watchwords. Furthermore, this strategy (if one can use such a word) entailed, at some point, the loss of its own self-consciousness. One of the strongest currents in Masse's rich and complex account follows the serialization of the *Amadis de Gaule* into the invention of actual serials—newspapers. From the "incremental newness" of an interminable romance, to the interminable romance of the incrementally new: there could be no better example of the period's invention of discovery.

Toward Hermeneutics

In the end, then, the chapters in this book offer a various collection of evidence for the hypothesis with which I began: that early-modern European culture offers us an hermeneutic scene in which discovery, as it were, was waiting to be invented. Organized around, but not limited to, the history of early-modern science, these essays meet in the busy contact zone between the empirical and the literary. To be sure, the arguments collected here are far from definitive; readers will think of many exceptions and confounds to the generalizations that a book like this one, almost inevitably, entails. Yet even to open a space for questioning within the modern and postmodern rule of discovery is to perform, I think, worthwhile work.

The techniques with which that work is done here are, for the most part, philological and historiographic. They contribute, if at all, to our understanding of the early-modern period. Ultimately, however, this contribution leads to theoretical reflection. It leads to hermeneutics in something like Gadamer's sense: to the understanding, if that be possible, of understanding itself.[27] There can be little

---

[27] See Gadamer, *Truth and Method*, pp. xi–xxxvi and *passim*; "The Universality of the Hermeneutic Problem," in Richard E. Palmer (ed.), *The Gadamer Reader* (Evanston, 2007), pp. 72–88; *The Enigma of Health*, trans. Jason Gaiger and Nicholas Walker (Stanford, 1996); Lewis Edwin Hahn (ed.), *The Philosophy of Hans-Georg Gadamer* (Chicago, 1997); and Jeff Malpas, Ulrich Arnswald and Jens Kertscher (eds), *Gadamer's Century: Essays in Honor of Hans-Georg Gadamer* (Cambridge, MA, 2002).

doubt that the hermeneutics of discovery, as the model of valid data-interpretation in modern natural science, controls and pre-determines what counts as valid data-interpretation generally, in the modern and postmodern intellectual culture that natural science dominates. Neither can there be any doubt—in my opinion—that the attempt to come to terms with these relevant aspects of the early-modern period necessitates a corollary attempt to come to terms with hermeneutics. This is not an optional or ancillary, but a core and crucial, part of the puzzle for early-modern studies. In this book's Afterword, I will try to make this theoretical case more fully. For now, though, it is time to turn to the period-based arguments that make up *The Invention of Discovery, 1500–1700*.

Chapter 1

"That full-sail voyage": Travel Narratives and Astronomical Discovery in Kepler and Galileo

Piers Brown

In the prefatory epistle to his *Astronomia nova* (1609), Johannes Kepler uses a series of related allegorical images to celebrate his discovery of the elliptical orbit of the planet Mars. He begins by announcing to his Imperial master, Rudolf II: "I am now at last exhibiting for the view of the public a most Noble Captive, who has been taken for a long time now through a difficult and strenuous war waged by me under the auspices of Your Majesty."[1] Kepler's description figures his book as a triumph in which, "riding in the triumphal car," he "will display the remaining glories of our captive that are particularly known to [him], as well as all the aspects of the war, both in its waging and in its conclusion."[2] Kepler goes on to describe how Mars was brought to bay:

> whenever he [i.e. Mars] was driven or fled from one castle, he repaired to another, all of which required different means to be conquered, and none of which was connected to the rest by an easy path—either rivers lay in the way, or brambles impeded the attack, but most of the time the route was unknown.[3]

While the sites of difficulty in his calculations are described as castles that must be besieged and overcome, and Mars, the hunt's quarry, provides an objective for the chase, the unknowability of the paths between these sites presents a problem that must be addressed.

Kepler's imagery provides an important comparison to William Eamon's work on the *venatio*, or hunt, as a scientific metaphor in the early-modern period. For Eamon, the *venatio* embodied a new, objective attempt to uncover the secrets

---

[1] Johannes Kepler, *New Astronomy*, trans. William H. Donahue (Cambridge, 1992), p. 30. See also Bruce Stephenson, *Kepler's Physical Astronomy* (New York, 1987); and James R. Voelkel, *The Composition of Kepler's* Astronomia nova (Princeton, 2001).

[2] Kepler, *New Astronomy*, p. 31.

[3] *Ibid.*, pp. 33–4.

of nature.[4] Eamon connects this new epistemological attitude to early-modern narratives of travel and exploration. Yet the *venatio* and the travel narrative—implied by Kepler's dedication to Rudolph (as cited above)—would seem to suggest quite distinct tropes for seeing the world. The *venatio* is primarily goal-oriented, deliberately seeking out the traces of its quarry (the facts). On exactly that basis, however, the *venatio* is vulnerable to the epistemological paradox of pre-cognition, or foreknowledge. Only if one knows, to some extent, what facts are out there can one know how, and where, to hunt for them. The travel narrative, by contrast, entails no such paradox. Navigational exploration can function randomly or haphazardly, being no more interested in its putative destination than in the places, peoples, and objects that it encounters along the way. Precisely by enabling a mode of *accidental* discovery, in other words, travel and exploration are superior to the hunt as tropes for scientific objectivity.

In this chapter, I will examine the use of travel narrative (as opposed to *venatio*) in Kepler's *Astronomia nova* (1609) and Galileo's *Sidereus nuncius* (1610). I will argue that the trope offers not only an organizational frame for historical accounts of exploration, but also an hermeneutics of discovery that focuses on error as beneficial precisely because of the accidental discoveries it produces. To this end, I will discuss the importance of travel narratives as a form of early-modern knowledge-making; examine Kepler's description of travel narrative as a system for historical narrative in his account of the mathematical exploration of Mars's orbit; and consider Kepler's insistence on the importance of difficulty as a vital component of the process of discovery. Finally, I will turn to the *Sidereus nuncius* and discuss its narrative of accidental discovery in the context of Kepler's travel metaphor.

"Although we by no means become Argonauts": Narrative Organization and Mathematical Exploration in Kepler's *Astronomia nova*

During the European age of expansion, travel produced not only new trade goods and accounts of foreign places and peoples, but also a paradigmatic image of the acquisition of knowledge. Old forms and ways of knowledge shifted and transformed in response to the influx of new geographical and anthropological information.[5] Francis Bacon signals the importance of long-distance networks of trade to early-modern knowledge production in his depiction of the idealized

---

[4]   William Eamon, *Science and the Secrets of Nature: Books of Secrets in Medieval and Early Modern Culture* (Princeton, 1994), pp. 269–300; quotation at p. 271. On books of secrets, see also Paula Findlen, *Possessing Nature: Museums, Collecting, and Scientific Culture in Early Modern Italy* (Berkeley, 1994); and Allison Kavey, *Books of Secrets: Natural Philosophy in England, 1550–1600* (Urbana, 2007).

[5]   See Anthony Grafton, with April Shelford and Nancy Siraisi, *New Worlds, Ancient Texts: The Power of Tradition and the Shock of Discovery* (Cambridge, MA and London,

scientific community in the utopian *New Atlantis* (1625). A significant component of this research community are the "Merchants of Light": scientific intelligencers who travel incognito, seeking out and gathering together information on technological developments and natural particulars from all corners of the world:

> For the several employments and offices of our fellows; we have twelve that sail into foreign countries, under the names of other nations (for our own we conceal); who bring us the books, and abstracts, and patterns of experiments of all other parts. These we call the Merchants of Light.[6]

Travel narratives were an important early-modern method of transmitting knowledge gained during the process of exploration. As Ann Blair has pointed out, the order of a work posed a serious problem for natural philosophy in the sixteenth and seventeenth centuries.[7] While earlier genres, such as the *navigatio*, were comparatively disorganized, the late sixteenth century saw the development of the "relation"—a broad tradition that included the *Relazioni* of Venetian ambassadors, and the Jesuit relations—along with a formal theory of travel writing, the *Ars apodemica*.[8] As a result, the genre of the travel narrative came to provide a framework for gathering "facts" into natural histories.[9] This framework also proved to be strikingly useful for other presentations of scientific knowledge.

In the *Astronomia nova*, Kepler uses specifically geographical, not cosmological, thinking to explain the very organization of his astronomical work. He compares his account to a travel narrative, in which the incidents of the voyage, as much as the destination, are likely to be of interest to the reader:

1992); and Harold Cook, *Matters of Exchange: Commerce, Medicine, and Science in the Dutch Golden Age* (New Haven and London, 2007).

[6]   Francis Bacon, *The Major Works*, ed. Brian Vickers (Oxford, 1996), p. 486.

[7]   See Ann Blair, *Theater of Nature: Jean Bodin and Renaissance Science* (Princeton, 1997), pp. 30–40, 49–81.

[8]   See Justin Stagl, *The Age of Curiosity: The Theory of Travel, 1550–1800* (Chur, 1995), pp. 47–94. On the relation, see Filippo de Vivo, *Information and Communication in Venice: Rethinking Early Modern Politics* (Oxford, 2007); and Steven J. Harris, "Mapping Jesuit Science: The Role of Travel in the Geography of Knowledge," in John W. O'Malley, Gauvin Alexander Bailey, Steven J. Harris, and T. Frank Kennedy (eds), *The Jesuits: Cultures, Science, and the Arts, 1540–1773* (Toronto, 1999), pp. 212–40.

[9]   See Barbara Shapiro, *A Culture of Fact: England, 1550–1720* (Ithaca and London, 2000), pp. 63–85; Marjorie Swann, *Curiosities and Texts* (Philadelphia, 2001), pp. 97–148; Lorraine Daston, "Baconian Facts, Academic Civility, and the Pre-History of Objectivity," *Annals of Scholarship*, 8 (1991): 337–63; Peter Dear, *Discipline and Experience: The Mathematical Way in the Scientific Revolution* (Chicago, 1995); and Mary Poovey, *A History of the Modern Fact: Problems of Knowledge in the Sciences of Wealth and Society* (Chicago: University of Chicago Press, 1998).

Here it is a question not only of leading the reader to an understanding of the subject matter in the easiest way, but also, chiefly, of the arguments, meanderings [*ambagibus*], or even chance occurrences by which I the author first came upon that understanding. Thus, in telling of Christopher Columbus, Magellan, and of the Portuguese, we do not simply ignore the errors [*errores*] by which the first opened up America, the second, the China Sea, and the last, the coast of Africa; rather, we would not wish them omitted, which would indeed be to deprive ourselves of an enormous pleasure in reading. So likewise, I would not have it ascribed to me as a fault that with the same concern for the reader I have followed this same course in the present work. For although we by no means become Argonauts by reading of their exploits, the difficulties and thorns of my discoveries infest the very reading—a fate common to all mathematical books. Nevertheless, since we are human beings who take delight in various things, there will appear some who, having overcome the difficulties of perception, and having placed before their eyes all at once this entire sequence of discoveries, will be inundated with a very great sense of pleasure.[10]

Kepler's comparison of his work with travel narratives draws attention to the genre's effectiveness as a narrative and organizational form. He suggests that the *Astronomia nova*, like the accounts of Christopher Columbus, Magellan, and the Portuguese—even the Argonauts—will hold the attention of its readers because of the surprises of its meandering form and the pleasures of vicarious discovery. Only once the work has been read fully will its whole shape be understood, laid out "all at once" as if in an idealized theater of knowledge.[11] Yet this description of the errant narrative of the *Astronomia nova* directly prefaces, and purports to explain, the chapter summaries by which Kepler lays out the contents of his book.

During the early-modern period, knowledge of the structure of the universe—the earth and the heavens—was, of course, subsumed under the discipline of cosmology. In this context, it might seem straightforward to map geographical travel narratives onto the astronomical exploration of the heavens, by metaphorically applying one to the other. However, formally speaking, astronomy was not part of cosmology, but instead preserved a separate character as a mathematical discipline intent on prediction. As such, astronomy's mathematical descriptions "saved the appearances," while abdicating to cosmology the responsibility for describing the form of the heavens. Only after Copernicus's claims to discover the heavens by astronomical calculation could a connection be made between active geographical exploration and seemingly-passive astronomical observation. This is consistent with the rhetorical and imaginative problem that Kepler confronts.

---

[10]   Kepler, *New Astronomy*, pp. 78–9.

[11]   On the theater as a metaphor for the organization of knowledge, see Blair, *The Theater of Nature*; and William N. West, *Theatres and Encyclopedias in Early Modern Europe* (Cambridge, 2002).

At issue here is Kepler's problem of how to organize the discussion of his astronomical discoveries, given the new methodological approach of his work. The extent of his difficulties is apparent in his battery of prefatory material. The main body of the work is preceded by a series of different introductions: the triumphal dedicatory letter to his patron, the Emperor Rudolf II (which I have already described); a series of dedicatory poems; an introduction for the "physicists," in which he presents arguments in favor of the heliocentric system specifically directed at natural philosophers; and, finally, an "elucidating introduction" offering a series of synoptic tables, chapter summaries, and an index of terms and authors.[12] This elaborate apparatus is necessary because, as Kepler complains, "it is extremely hard these days to write mathematical books, especially astronomical ones."[13] His intention is contrary to the conventions of astronomical writing: it "is not chiefly to explain the celestial motions. Nor yet is it to teach the reader, to lead him from self-evident beginnings to conclusions, as Ptolemy did as much as he could."[14] In other words, astronomical descriptions are usually organized either in terms of physics—based on the form of the thing described; or mathematics— built up from axioms like Euclidean geometry. While the first method is obviously not appropriate, because the subject of Kepler's book is not primarily physics, the second poses a problem: "For unless one maintains the truly rigorous sequence of proposition, construction, demonstration, and conclusion, the book will not be mathematical; but," he warns, "maintaining that sequence makes the reading tiresome."[15]

Accordingly, Kepler turns to the genre of *historia*, in which matter is organized in a chronological rather than a logical fashion.[16] Mathematically, the *Astronomia nova* is a series of iterative approximations and refutations based upon Tycho Brahe's data, by which Kepler eventually produces the conclusion that Mars's orbit of the Sun is elliptical in shape.[17] But "along with the former methods," Kepler "mingle[s] the third, familiar to the orators; that is, an historical presentation of my discoveries."[18] The *Astronomia nova* is, Kepler suggests, more a journal of his work than a typical methodized text. Yet despite Kepler's claims, the *Astronomia nova* was not a strictly or minutely historical account of his investigations.[19] As

---

[12]   Kepler, *New Astronomy*, p. 46.

[13]   *Ibid.*, p. 44.

[14]   *Ibid.*, p. 78.

[15]   *Ibid.*, p. 45.

[16]   See Donald R. Kelly, "Between History and System," in Gianna Pomata and Nancy G. Siraisi (eds), *Historia: Empiricism and Erudition in Early Modern Europe* (Cambridge, MA and London, 2005), pp. 211–37.

[17]   See Stephenson, *Kepler's Physical Astronomy*; and Voelkel.

[18]   Kepler, *New Astronomy*, p. 78.

[19]   See Owen Gingerich, "The Computer versus Kepler," *American Scientist*, 52/2 (1964): 218–26; and "Kepler's Treatment of Redundant Observations or, the Computer versus Kepler Revisited," in F. Krafft, K. Meyer, and B. Sticker (eds), *Proceedings of the*

William Donahue notes, Kepler does not "claim to present such a history: he says only that he is mingling some history with the theoretical and didactic matter of the book."[20] Kepler "describe[s] only so much of that labour of four years as will pertain to [his] methodical enquiry."[21] The resulting mixed form aimed to provide both a rigorous geometrical argument and a persuasive explanatory method.

However, the combination of the two ran the risk of confusing the reader. It is for this reason that travel narrative is used to explain the synoptic table and its accompanying chapter summaries. Kepler explains that

> the synopsis will not be of equal assistance to all. There will be those to whom this table (which I present as a thread leading through the labyrinth of the work) will appear more tangled than the Gordian Knot. For their sake, therefore, there are many points that should be brought together here at the beginning which are present bit by bit throughout the work, and are not so easy to attend to in passing.[22]

Although he claims that the reader will find his work to be "well ordered through this method," that order comes out of the very wandering path of his investigations.[23]

Although earth-bound astronomy was limited to observation, as the chapter summaries show, Kepler's process of mathematical approximation functions as a metaphorical "exploration" of the orbit of the planet Mars, in which he attempts to trace the movement of the planet, to follow its path through the stars precisely. The narrative structure of the work follows Kepler's attempts to determine its course as they go off track, each part of the work returning to a new starting point with new assumptions. For instance:

> When I was on this same path and was confronted with this equivocal fork in the road (in chapters 19 and 20 above), and the observations (most faithful guides) were seen to be at war with observations, the thought occurred to me to alter completely the way the path was set out, using the method which follows.[24]

Kepler's investigation is a process of exploration that turns back on itself, going through similar attempts to approximate as successive hypotheses break down. Each of the five parts of the main narrative begins again, based upon new premises

---

Internationales Kepler-Symposium, Weil der Stadt 1971 (Hildesheim, 1973), pp. 307–14. Both are reprinted in Gingerich, *The Eye of Heaven: Ptolemy, Copernicus, Kepler* (New York, 1993). See also Donahue, "Translator's Introduction," in Kepler, *New Astronomy*, pp. 1–19.

20   Kepler, *New Astronomy*, p. 79, n. 1.
21   *Ibid.*, p. 187.
22   *Ibid.*, pp. 45–6.
23   *Ibid.*, p. 79.
24   *Ibid.*, p. 305.

discovered in the previous stage. This approach transforms dead ends, encountered during attempts to fit different mathematical models to observations, into new points of departure.

Kepler's interest in historical narrative and its accidents is apparent throughout the *Astronomia nova*. Thus, for instance, in chapter seven, Kepler presents an account of the circumstances by which he first came to study astronomy, joining together his interest in the structure of the cosmos and the "invisible agency" of fate:

> It is true that a divine voice, which enjoins humans to study astronomy, is expressed in the world itself, not in words or syllables, but in things themselves and in the conformity of the human intellect and senses with the sequence of celestial bodies and their dispositions. Nevertheless, there is also a kind of fate, by whose invisible agency various individuals are given to take up various arts, which makes them certain that, just as they are part of the work of creation, they likewise also partake to some extent in divine providence.[25]

Kepler goes on to explain that, while he was seduced by "the sweetness of philosophy," he had "no special interest in astronomy." It merely so happened that the first job "to offer itself was an astronomical position."[26] As a result of his work on astronomy, he wrote the *Mysterium cosmographicum* (1599) and was driven by his "own ardour to seek, through a reworking of astronomy, whether my discovery [the speculations in the *Mysterium*] would stand comparison with observations made with perfect accuracy."[27] It was because of this interest in observations that he wrote to Brahe and went on to work with him. Moreover, his interest in the planet Mars came about because it "happened by divine arrangement, that I arrived at the same time in which he [Brahe] was intent upon Mars," when he otherwise might have become interested in another of the planets.[28]

This account, with its emphasis on coincidence and happenstance, also bears a strong resemblance to the fortuitous arrival of Duracotus at Brahe's observatory on Hven in Kepler's *Somnium*, which itself takes the form of a series of voyages. Duracotus is sold to a sea captain by his mother, but because of his sea-sickness is left on the island of Hven, where he comes into the service of Brahe. Here, he is driven by his particular circumstances—including his own mother's magical communions—to investigate not Mars, but rather the moon.[29] It is the very circumstantiality of these stories and their unexpected turns in response to difficulties that Kepler hoped would produce pleasure in the reader.

---

[25]   *Ibid.*, p. 183.

[26]   *Ibid.*

[27]   *Ibid.*, p. 184.

[28]   *Ibid.*, p. 185.

[29]   Johannes Kepler, *Somnium: The Dream, or Posthumous Work on Lunar Astronomy*, trans. Edward Rosen (Madison, 1967).

"The thorns in my discoveries": Difficulty and the Accidents of Discovery

The deliberately haphazard chronological argument of the *Astronomia nova* seems, at first glance, to contradict Fernand Hallyn's account of the poetics of Kepler's astronomy.[30] Hallyn draws attention to the importance of "mannerist aesthetics" in Kepler's works, particularly the geometrical and musical speculations of the *Mysterium cosmographicum* (1596) and the *Harmonicae mundi* (1619). However, the wandering path of Kepler's argument, and the emphasis on order and harmony in his conception of God's order, are not contradictory; but rather meet in his focus on the perplexing difficulties that he encountered during his investigations.

The dedicatory epistle to the *Astronomia nova* emphasizes the strangeness of Mars's orbit as an incitement to astronomical exploration:

> the eternal Architect of this world, and the Father of the Heavens and Humans in common, Jupiter located [Mars] in the front lines of the visible bodies ... so that he might raise human minds, lulled to sleep by a deep somnolence, from the slanderous reproach of idleness and ignorance, arouse them to venture forth, and provoke them forcefully to carry out investigations in the heavens for the praise of their Creator.[31]

Kepler's description of this incitement to investigation aligns with a widespread understanding of the importance of wonder and curiosity to early-modern knowledge-making.[32] These passions were aroused by encounters with strange objects, places, and peoples—what Bacon would call the "heteroclites or irregvlars of nature"—that both attracted attention and impeded understanding.[33] These "diversions" and "digressions" from the course of nature draw attention in the same way as the *errores*, the wanderings and mistakes, of Kepler's narrative. These accidents, as Donahue has suggested, were a crucial part of Kepler's innovative

---

[30]   Fernand Hallyn, *The Poetic Structure of the World: Copernicus and Kepler*, trans. Donald M. Leslie (New York, 1990). See also Gérard Simon, "Analogies and Metaphors in Kepler," in Fernand Hallyn (ed.), *Metaphor and Analogy in the Sciences* (Dordrecht, 2000), pp. 71–82; J.V. Field, *Kepler's Geometrical Cosmology* (Chicago, 1988); and Bruce Stephenson, *The Music of the Heavens: Kepler's Harmonic Astronomy* (Princeton, 1994).

[31]   Kepler, *New Astronomy*, pp. 31–2. Also quoted in Voelkel, p. 219; see n. 43 for Voelkel's translation.

[32]   See Stagl; Swann; Lorraine Daston and Katherine Park, *Wonders and the Order of Nature* (New York, 1998); Mary Baine Campbell, *Wonder and Science: Imagining Worlds in Early Modern Europe* (Ithaca, 1999); and Pamela H. Smith and Paula Findlen (eds), *Merchants and Marvels: Commerce, Science and Art in Early Modern Europe* (New York, 2002).

[33]   Francis Bacon, *The Advancement of Learning*, ed. Michael Kiernan (Oxford, 2000), p. 63.

approach to discovery.[34] In his introduction, Kepler notes that following the path of Mars was not easy: "either rivers lay in the way, or brambles [*sentibus*] impeded the attack, but most of the time the route was unknown."[35] In the explanation of the synopsis, he complains that in his mathematical work, "the difficulties and thorns [*spinae*] of my discoveries infest the very reading."[36] These thorny difficulties suggest a further connection with travel narratives. The "thorns" which infest his work not only suggest the difficulties of the process of exploration, but also the discoveries made on those voyages—the foreign plants that might be brought back from the new world and subsequently domesticated.[37]

The interest of the work, then, is not just in its conclusions, but also in the use that can be made of its difficulties in later work. Mathematical difficulties, like the plants encountered by explorers of the new world, could become objects of study in their own right. Thus Kepler emphasizes this orientation metaphorically, as in his deployment of an image of winnowing to describe the process of separating out useful results from his work:

> How small a heap of grain we have gathered from this threshing! But you also see what a huge cloud of husks there is now. They ought to have been hauled back to the beginning of ch. 48, since before I investigated the arcs of the oval path I would have dealt with them. But for the sake of bringing light, they ought to be winnowed. Besides, we might end up finding a few useful grains.[38]

Plants that seem to be obstacles are found to produce valuable seeds, bark or roots. Similarly, Kepler's mathematical discoveries might, like the facts in real travel narratives, be plucked out of the book and reused in other settings. As such, there is a sense in which we can see his text as a resource for further astronomical or mathematical work.

Although Kepler mentions these thorny impediments only in a couple of places in the *Astronomia nova*, they reappear in his illustrations: many of the complex diagrams illustrating the hypothetical relationships between the planets are surrounded by printed woodcut ornaments of flowers and plants. Decorative ornaments, particularly vines and plants, were in fact sometimes used in the period to protect delicate woodcuts of geometrical diagrams.[39] However, in the *Astronomia*

---

[34]   Kepler, *New Astronomy*, p. 79.

[35]   *Ibid.*, pp. 33–4.

[36]   *Ibid.*, p. 79.

[37]   See Cook, pp. 304–38; Londa Schiebinger, *Plants and Empire: Bioprospecting in the Atlantic World* (Cambridge, MA, 2004); and Londa Schiebinger and Claudia Swann (eds), *Colonial Botany: Science, Commerce, and Politics in the Early Modern World* (Philadelphia, 2005).

[38]   Kepler, *New Astronomy*, p. 495.

[39]   This may be achieved either by carving ornaments into the woodblock or by inserting metal ornaments. In the case of the *Astronomia nova*, it appears to be the former. I am

nova the practice thematically connects itself with Kepler's metaphor and, as a result, draws attention to the complexity of the diagrams as sites in his work. In this context, Kepler's mathematical thorns can also be read as a counterpart to the poetic and rhetorical flowers that humanists gathered during their reading, often marked in the margins with a trefoil symbol. Ann Blair has suggested that the juxtaposition of different material in the commonplace book acted as a stimulus to early-modern natural philosophy.[40] In Kepler's work we see both the results of this practice of poetic gathering, and the use of poetic allusion to draw attention to cruxes in his argument.

These different modes come together in chapter 58, Kepler's account of the *via buccosia* or puff-cheeked orbit. He begins the chapter with a quotation from Virgil (*Eclogues*, 3.64):

Galatea seeks me mischievously [*malum*], the lusty wench:

She flees the willows, but hopes I'll see her first.

It is perfectly fitting that I borrow Vergil's voice to sing this about Nature. For the closer the approach to her, the more petulant her games become, and the more she again and again sneaks out of the seeker's grasp just when he is about to seize her through some circuitous route. Nevertheless, she never ceases to invite me to seize her, as though delighting in my mistakes.[41]

As Donahue points out, Virgil's *malum* is also the Latin for "apple," implicitly suggesting the hypothesis of a puff-cheeked orbit. As James Voelkel has argued, this passage, perhaps more than any other, shows that Kepler's organization of the text is both persuasive and explanatory. Voelkel suggests that Kepler included the example of the puff-cheeked path in the *Astronomia nova* specifically because, when discussing his investigations with David Fabricius, the omission of this erroneous step had confused his correspondent and led to disagreements.[42] Kepler's decision to allude to this particular fruit of his investigations by means

---

indebted to Roy Laird and Randall McLeod for these observations. See for comparison the use of ornament in the diagrams of Pappus of Alexandria, *Pappi Alexandrini Mathematicae collectiones*, trans. Federico Commandino (Pisa, 1588). On the use of bearing type in printing, see Random Cloud [Randall McLeod], "Where Angels Fear to Read," in Joe Bray, Miriam Handley, and Anne C. Henry (eds), *Ma(r)king the Text* (Aldershot, 2000), pp. 144–92.

[40]   Ann Blair, "Humanist Methods in Natural Philosophy: The Commonplace Book," *Journal of the History of Ideas*, 53/4 (1992): 541–51.

[41]   Kepler, *New Astronomy*, p. 573. See also Voelkel, p. 245; and Pierre Hadot, *The Veil of Isis: An Essay on the History of the Idea of Nature*, trans. Michael Chase (Cambridge, MA, 2006).

[42]   Voelkel, pp. 193–210.

of poetic quotation draws yet more attention to the passage. Moreover, it raises the possibility of reading the image allegorically, connecting its mischievousness with the false knowledge that Eve gained by eating the apple in the Garden of Eden (a context in which the pun on *malum* was very common). The reader could connect this episode with other botanical imagery in Kepler's text, as well as with its explicit imagery of pursuit.

Sea Voyages and the Movement of the Earth

Kepler marks the significant transition between parts three and four of his text with a quotation from Ovid: "Part of what has been begun, part of the work is finished: The anchor is cast; here let the craft lie."[43] By marking the pause between arguments by a shift into the poetic register, Kepler reminds his readers of his organizational metaphor of the travel narrative, as they rest for a moment in preparation for the next stage. In doing so, however, he also returns to the theme of the voyage, which plays an important role in Kepler's attempts to address concerns about the instability of a moving earth.

Although the *Astronomia nova* substitutes elliptical orbits for the regular circuits of Copernicus, it is based on Copernican foundations. Therefore, Kepler was confronted by the need to address scriptural authority, with its accounts of the centrality and immobility of the earth.[44] Kepler responds to these "objections concerning the dissent of holy scriptures, and its authority" by distinguishing between literal and accommodationist views of scripture. He argues that

> since we acquire most of our information, both in quality and quantity, through the sense of sight, it is impossible for us to abstract our speech from this ocular sense. Thus, many times each day we speak in accordance with the sense of sight, although we are quite certain that the truth of the matter is otherwise. This verse of Vergil furnishes an example: "We are carried from the port, and the land and cities recede." Thus, when we emerge from the narrow part of some valley, we say that a great plain is opening itself out before us.[45]

---

[43]   Kepler, *New Astronomy*, p. 427.

[44]   See Robert S. Westman, "The Copernicans and the Churches," in David C. Lindberg and Ronald L. Numbers (eds), *God and Nature: Historical Essays on the Encounter between Christianity and Science* (Berkeley, 1986), pp. 76–113; Jean Deitz Moss, *Novelties in the Heavens: Rhetoric and Science in the Copernican Controversy* (Chicago, 1993), pp. 129–32; and James M. Lattis, *Between Copernicus and Galileo: Christoph Clavius and the Collapse of Ptolemaic Cosmology* (Chicago, 1994), pp. 120–26.

[45]   The quotation is from Virgil, *Aeneid* III. 72. Compare Copernicus's use of the same quotation: "when a ship is floating calmly along, the sailors see its motion mirrored in everything outside, while on the other hand they suppose that they are stationary, together with everything on board. In the same way, the motion of the earth can unquestionably

In this passage, Kepler reuses a poetic description from Virgil that Copernicus used to explain the same problem of the apparent motion of the heavens. Both invoke poetry to show the validity of metaphorical language, in contrast to literal readings of scripture. As with the quotation from Virgil discussed above, Kepler's poetic quotation draws attention to his point. It also suggests the importance of voyages as a metaphor for the new conception of the earth's place in the heavens.

Kepler pursues the metaphor in the same section of the *Astronomia nova*, in response to objections concerning the immensity of the heavens:

> In the same place, you will find that full-sail voyage [*plenis velis nauigatum*] along the world's immense orbit, which is usually held to be unnatural, in objection to Copernicus. There it is demonstrated to be well-proportioned, and that, on the contrary, the speed of the heavens would become ill-proportioned and unnatural were the earth ordered to remain quite motionless in its place.[46]

Here, an apparent difficulty, once resolved, itself offers a solution to other problems. Copernicus's reconception of the universe as sun-centered is expressed by the image of earth as a ship and its orbit as a "full-sail voyage of discovery." Not only can the apparent motion of the heavens be seen as the result of the earth's triple motion, but the problem of the vast speed of the heavens is also resolved. This freeing up of the earth leads to speculation about the possibilities inherent in the observation of the universe from a mobile, as opposed to a static, vantage point.

In his response to Galileo's *Sidereus nuncius* (1610), the *Dissertatio cum nuntio sidereo* (1611), Kepler suggests that conceiving of the moving earth as a boat allows the observer to understand it as a mobile platform from which the processes of surveying and triangulation can be undertaken:

> as I said in the "Optics," in the interests of that contemplation for which man was created, and adorned and equipped with eyes, he could not remain at rest in the center. On the contrary, he must make an annual journey on this boat, which is our earth, to perform his observations. So surveyors, in measuring inaccessible objects, move from place to place for the purpose of obtaining from the distance between their positions an accurate base line for the triangulation.[47]

Kepler's description of astronomical observation in terms of surveying suggests how Copernicus's hypothesis offers a powerful new way of understanding man's place in the universe. In this vision, it is precisely the movement of the earth that provides new sights, transforming astronomy from a static process of observation

---

produce the impression that the entire universe is rotating." Nicholas Copernicus, *On the Revolutions*, trans. Edward Rosen (Baltimore, 1992), p. 16.

[46]   Kepler, *New Astronomy*, p. 59.

[47]   Johannes Kepler, *Kepler's Conversation with the Starry Messenger*, trans. Edward Rosen (New York, 1965), p. 45.

into a voyage of cosmographical exploration. The very compilation of celestial observations becomes a travel narrative—a series of sights, seen in different places, that may reconfigure themselves in surprising ways when read in order.

Starry Messages: Galileo's Journal of His Observations

In conclusion, I want to turn to an even more famous publication than Kepler's, from the year following the *Astronomia nova*: Galileo's account of his observations with the telescope, the *Sidereus nuncius* (1610). Galileo, like Kepler, emphasizes the *accidental* character of his discoveries. He enhances the credibility of his observations by portraying them as discovery-driven rather than goal-driven. The heavens that he describes seeing through the telescope are filled with unexpected phenomena that attract his interest. This rhetoric of exploration and surprise is clearest in Galileo's discovery of the Medicean stars, as he calls the moons of Jupiter. He reports that, on the seventh of January, 1610, he observed Jupiter for the first time with his improved telescope and saw that

> three little stars were positioned near him—small but yet very bright. Although I believed them to be among the number of fixed stars, they nevertheless intrigued me because they appeared to be arranged along a straight line and parallel to the ecliptic, and to be brighter than others of equal size.[48]

Galileo's curiosity about these stars was rewarded, "when on the eighth, I returned to the same observation, guided by I know not what fate, [and] found a very different arrangement. For all three little stars were to the west of Jupiter."[49] He moved "from doubt to astonishment,"[50] and was impelled to complete a long series of observations, presented in his book, after which he concluded that the "little stars" must be planets orbiting Jupiter. The sequence represents, in miniature, an equivalent to Kepler's far more complicated discovery of the orbit of Mars.

The connection between the new astronomy and the European age of exploration is an established critical trope. William Shea, for instance, has argued that the reconceptualization of the universe central to the Copernican hypothesis, which transformed the planets from wandering stars into alternative worlds, enabled metaphorical connections to be made between the new exploration of the heavens and Columbus's discovery, and the subsequent exploration and colonization, of the new world.[51] Such arguments, however, place an emphasis on the ability

---

[48]   Galileo, *Sidereus Nuncius or the Starry Messenger*, trans. and intro. Albert Van Helden (Chicago, 1989), p. 64.

[49]   *Ibid.*, p. 65.

[50]   *Ibid.*, p. 66.

[51]   William Shea, "Looking at the Moon as another Earth: Terrestrial Analogies and Seventeenth-Century Telescopes," in Fernand Hallyn (ed.), *Metaphor and Analogy in the*

to identify seas, coastlines, or other geographical features; rather than on the organizational form and emphasis on accidental discovery that Kepler and Galileo associate with travel narratives in their work. Both Kepler and Galileo place an epistemological value on the delights of accidence, precisely as opening up new modes of mathematical investigation, and new observational habits. The rhetorical use of the travel report by these two giants of early-modern astronomy suggests a surprising construction, via pleasure and happenstance, of the factual sensibility that would be central to the natural philosophy of the scientific revolution.

Sciences (Dordrecht, 2000), pp. 83–103. On this trope in Kepler's work, see Mary Baine Campbell, "Alternative Planet: Kepler's *Somnium* (1634) and the New World," in Claire Jowitt and Diane Watt (eds), *The Arts of 17th-Century Science: Representations of the Natural World in European and North American Culture* (Aldershot, 2002), pp. 232–49.

Chapter 2

Francis Bacon and the
Divine Hierarchy of Nature

Steven Matthews

In Aphorism 70 of the *Novum Organum*, Sir Francis Bacon famously insisted that "experiments of light," or discovery, must precede "experiments of fruit," or those inventions which would follow. The rush to invention before discovery, for Bacon, was one reason for the lack of both knowledge and useful invention in ages past. Moreover, it confused the very divine order, since God established "light" on the first day of creation, with fruit following.[1] Through arguments like this one, Bacon significantly contributed to the Enlightenment binary of discovery over invention, discussed in the Introduction to this volume. Voltaire credited Bacon with "raising the scaffolding" by which the new experimental sciences had been built.[2] In very Enlightenment fashion, however, the French *philosophe* failed to acknowledge the religious implications and assumptions of his English predecessor's construction work.

Recent scholarship has reopened the once-closed question of Bacon's indebtedness to Christian tradition. It was long thought (or assumed) that Bacon sought merely to separate emergent science from religion. It is now apparent, however, that Bacon's innovative natural philosophy depends upon a certain theological system, and a unique interpretation of scripture and the church fathers.[3] Bacon's main goal in the "reform of learning" was not to establish "modern science," but to reform all fields of human knowledge. It stands to reason, accordingly, that the *sources* of Bacon's method for such a general reform of learning would draw upon traditions as diverse (to our eyes) as law, astronomy, alchemy, and theology. One apparently unique feature of Bacon's method was his use of "negative instances" to ascend to certainty in regard to the natural order.

---

[1]   James Spedding, Robert Leslie Ellis, and Douglas Dennon Heath (eds), *The Works of Francis Bacon* (14 vols, London, 1858–74), vol. I, p. 180 (hereafter cited in the form: *WFB*, I, p. 180).

[2]   Voltaire, *Letters on England* [*Lettres philosophiques*], trans. Leonard Tancock (New York, 1980), pp. 57–61.

[3]   See, among others: Stephen McKnight, *The Religious Foundations of Francis Bacon's Thought* (Columbia, 2005); Peter Harrison, *The Fall of Man and the Foundations of Science* (Cambridge, 2007); and Steven Matthews, *Theology and Science in the Thought of Francis Bacon* (Aldershot, 2008).

Scholars have called this hermeneutic method "eliminative induction," and have noted the importance Bacon attaches to it; but they have cast serious doubts over its natural-philosophical efficacy. This chapter will argue that Bacon's negative hermeneutics actually projects a theological, rather than an empirical, framework for the discovery of knowledge. To wit, eliminative induction correlates natural philosophy with the hierarchical cosmology of Pseudo-Dionysius.[4]

Baconian Negativity

Eliminative induction is the central process of Bacon's method of inquiry into nature.[5] By progressively eliminating that which a thing (or principle, or quality) is not, Bacon contends, one can come to an understanding of what it truly is. Like an alchemist performing repeated separations over the fire, true induction uses the mind (which Bacon labels a "divine fire" [*ignem divinum*]) to perform the separations necessary to understand what something *is*.[6] Bacon demonstrates via the example of heat. The first step is to gather a list of instances where heat occurs: from fire, to the rays of the sun, to the heat found in animal bodies. After gathering these into "Tables of Presentation," the mind can eliminate those which do not apply to heat in general, but only to specific instances, thereby isolating what "heat" is apart from the accidents of its manifestation:

1. On account of the rays of the sun, reject the nature of the elements.
2. On account of common fire, and chiefly subterraneous fires (which are most remote and most completely separate from the rays of heavenly bodies), reject the nature of heavenly bodies.[7]

---

[4]   Michael McCanles has linked Bacon to the Pseudo-Dionysian *via negativa*, but primarily by way of articulating Bacon's debt to Ockham, in terms of a generalized medieval mysticism: see McCanles, "The New Science and the *Via Negativa*: A Mystical Source for Baconian Empiricism," in Julie Robin Solomon and Catherine Gimelli Martin (eds), *Francis Bacon and the Refiguring of Early-Modern Thought* (Aldershot, 2005), pp. 45–68. Feisal G. Mohamed has recently examined the heritage of Pseudo-Dionysian angelology in English Renaissance literature and related discourses (such as episcopacy), but without reference to empiricism, or to Bacon. Mohamed establishes a carefully documented narrative of the *decline* of Dionysius in the English Reformation which closely parallels the summary here. See Mohamed, *In the Anteroom of Divinity: The Reformation of the Angels from Colet to Milton* (Toronto, 2008).

[5]   Stephen Gaukroger, *Francis Bacon and the Transformation of Early-Modern Philosophy* (Cambridge, 2001), p. 138. See also Brian Vickers, "Francis Bacon and the Progress of Knowledge," *Journal of the History of Ideas*, 53/3 (1992): p. 499.

[6]   *WFB*, IV, pp. 145–6; and for the Latin, *WFB*, I, p. 257.

[7]   *WFB*, IV, p. 147. Latin, *WFB*, I, p. 259.

In this manner the divine fire of the mind can "burn off" all that does not apply to the essential nature of "heat."

Once the elimination is completed, Bacon prescribes a process of analytical reflection, which he terms "the Indulgence of the Understanding, the Commencement of Interpretation, or the first vintage."[8] What is left at the end of the process is an affirmative statement about heat, a "lawlike" description and definition, which is the phenomenon's essential "form."[9] This hypothetical definition may then be further tested and refined.[10] Although Bacon was only using heat as an example, and he admitted that his investigation had not been thorough, we may still be impressed by Bacon's conclusion that heat was (according to "form") nothing other than a variety of motion.[11] It would require another two and a half centuries, which included the rise and fall of "caloric theory," for mainline science to establish the same conclusion.

Nonetheless, the process via which Bacon arrives at his result is problematic. Stephen Gaukroger has argued that, although Bacon regarded eliminative induction as the core of his method, its value is limited in determining causes. Other elements of Bacon's method are actually what make it work:

> In sum, it is difficult to find a case where eliminative induction does real work, where the other factors are not the crucial ones in the process. Moreover, it seems particularly ill-suited to discovering the material constituents or causes of macroscopic phenomena. Bacon has elaborated something which is useful … but its credentials as a method of discovery, as opposed to simply an aid, are quite impossible to establish.[12]

Yet for Bacon, the "credentials" of his method were not intrinsic. They were not established by the method's effectiveness, but derived from his reverence for the church fathers and Christian antiquity. In particular, Bacon's use of "negative instances" is an adaptation of the Pseudo-Dionysian *via negativa*.

Also known as negative theology, this is the process of coming to know the divine through negative statements, necessitated by God's absolute transcendence. Vladimir Lossky provides a concise explanation:

> Dionysius distinguishes two possible theological ways. One—that of cataphatic or positive theology—proceeds by affirmations; the other—apophatic or negative

---

8   *WFB*, IV, p. 149. Latin, *WFB*, I, p. 261.

9   There has been much debate over Bacon's use of "form." For its "lawlike" nature see Gaukroger, *Francis Bacon*, pp. 140–41. Compare Bacon's *Novum Organum* Book II, in *WFB*, I, pp. 257–8.

10   On the "first vintage" in Bacon's method see Lisa Jardine, *Francis Bacon: Discovery and the Art of Discourse* (Cambridge, 1974), pp. 127–8.

11   *WFB*, I, p. 266.

12   Gaukroger, *Francis Bacon*, pp. 152–3.

theology—by negations. The first leads us to some knowledge of God, but is an imperfect way. The perfect way, the only way which is fitting in regard to God, who is of His very nature unknowable, is the second—which leads us finally to total ignorance ... It is by unknowing (*agnosia*) that one may know Him who is above every possible object of knowledge. Proceeding by negations one ascends from the inferior degrees of being to the highest, by progressively setting aside all that can be known, in order to draw near to the Unknown in the darkness of absolute ignorance.[13]

The "ignorance" which is praised by Dionysius is not anti-intellectualism, but the result of a supreme act of the intellect. Knowledge of the created order is required. God, precisely because he is the originator of all forms of knowledge, must necessarily be beyond knowing. It is only through the intellectual activity of knowing creation at ever higher levels that the mind can ascend beyond the knowable to the recognition that God truly transcends all. As the Dionysian author wrote of the divine in the tract known as *Mystical Theology*:

There is no speaking of it, nor name nor knowledge of it. Darkness and light, error and truth—it is none of these. It is beyond assertion and denial. We make assertions and denials of what is next to it, but never of it, for it is both beyond every assertion, being the perfect and unique cause of all things, and, by virtue of its preeminently simple and absolute nature, free of every limitation, beyond every limitation, it is also beyond every denial.[14]

Since all of creation was arranged hierarchically, as on a ladder, the *cataphatic* and *apophatic* ways are described in terms of "descent" and "ascent" respectively: "When we made assertions we began with the first things, moved down through intermediate terms until we reached the last things. But now as we climb from the

---

[13]   Vladimir Lossky, *The Mystical Theology of the Eastern Church* (Crestwood, 1976), p. 25.

[14]   Pseudo-Dionysius the Areopagite, *The Complete Works*, trans. Colm Luibheid (New York, 1987), p. 141. Bacon would have used one of roughly a dozen different Latin translations available in his day, but it is impossible to say which. Traversari's translation was available in England and was the one used by John Colet, according to Daniel T. Lochman, "*Divus Dionysius:* Authority, Self, and Society in John Colet's Reading of the *Ecclesiastical Hierarchy*," *Journal of the History of Ideas*, 68/1 (2007): p. 14. Traversari's translation varies widely from the others and is much more interpretive, veering from the more word-for-word approach which dominated translations of Pseudo-Dionysius since Erigena. For this essay the non-Traversari texts are preferred. They are remarkably similar and form a consensus interpretation of the original Greek. I have compared all of the available Latin texts, using the critical edition of Philippe Chevallier, *Dionysiaca; recuil donnant l'ensemble des traductions latines* (Bruges, 1937), hereafter *Dionysiaca*. Compare the Latin translations in *Dionysiaca*, pp. 600–602.

last things up to the most primary we deny all things."[15] Assertions (the positive way) are only valid if the "first things" are known. For that knowledge which must arise from the "last things," the observations of the created order, the negative way, or the way of "denial" [*ablationes*] is required.[16]

Pseudo-Dionysius in History

Using the corpus of Pseudo-Dionysius was not without problems in Bacon's day, not so much because of the false identity of the author, but because of its dependency upon Plato, and its association with the arguments of Roman Catholics. That the author was not the first-century colleague of the Apostle Paul, as the writings themselves claim, but a fifth-century contributor to the monophysite controversy, was of no great importance to Bacon, Lorenzo Valla, or others who believed that what was *in* the documents was essentially a correct exposition of spiritual truth.[17] The question of whether the author was *right* in his exposition, and not whether he was Dionysius, was also central for opponents of the corpus.

In the medieval West the *via negativa* and the writings of "the Areopagite" were influential, but they never held sway as they did in the Christian East. In part this is due to the powerful influence of Maximus the Confessor in the East, whose commentaries ensured that the writings of Pseudo-Dionysius were recognized as unquestionably orthodox. Erigena, by contrast, who introduced the writings to the Latin West, had the stigma of being labeled a heretic. Also, the scholasticism which dominated the late medieval West had in many ways incorporated the Dionysian writings, thereby reducing the premium on the originals. More significantly, however, the positive scholasticism of Aquinas and Lombard tended to reduce the *via negativa* to an aspect of the theological enterprise, rather than its highest form. The Dionysian corpus, and its negative theology, remained strongest among the mystic movements and the more mystic scholastics (such as the Victorines and Bonaventure).[18] The greatest blow to the authority of the Dionysian texts in the West came through the polemics of the Reformation.

---

[15] Pseudo-Dionysius, *The Complete Works*, p. 138.

[16] *Dionysiaca*, p. 581.

[17] On the identity of the author of the texts I agree with Sarah Klitenic Wear and John Dillon who argue for the provenance of the corpus in the Monophysite debates after Chalcedon. See Wear and Dillon, *Dionysius the Areopagite and the Neoplatonist Tradition: Despoiling the Hellenes* (Aldershot, 2007), pp. 4–6. On the lack of interest in the problem of pseudepigraphy in classical Christianity see Bruce M. Metzger, "Forgeries and Canonical Pseudepigrapha," *Journal of Biblical Literature*, 91/1 (March, 1972): pp. 3–24. Our modern concern with the question of "forgery" is indeed modern.

[18] Paul Rorem, *Pseudo-Dionysius: A Commentary on the Texts and an Introduction to Their Influence* (Oxford, 1993), pp. 214–25.

Doubts about the authenticity of the Dionysian corpus surfaced in the West in the wake of Lorenzo Valla, who managed to question the authorship, without denying the authority, of the texts. With Luther and Calvin, a split developed along the lines of the Reformation itself.[19] In debates, Roman Catholics used the texts extensively to support the hierarchy of the Church and the significance of its sacraments. Therefore the Roman Catholics were more likely to argue for both authority and authenticity, although authority was paramount. Luther and Calvin dismissed the texts with prejudice, in no small part because of the use made of them by their opponents. It is significant that even in Protestant objections authorship was at best a secondary issue. As Karlfried Froehlich states, "for most Protestant Reformers, it was precisely the obvious Dionysian Platonism that became the focus of their unsympathetic assessment."[20] References to the texts among Protestants amounted to little more than disparaging them in attacks on their Romanist opponents. The exceptions that prove this rule are the writings of those such as Donne and Hooker, which may follow a Dionysian pattern, but do not make explicit connection to the Areopagite.[21] In the literature of Protestant England in Bacon's day, it is extremely rare to find a positive mention of the figure we now call Pseudo-Dionysius. Francis Bacon offers a notable exception.

Created Hierarchies

Pseudo-Dionysian thought resonates with Baconian natural philosophy. In the first book of *The Advancement of Learning*, Bacon explains that all of nature exists in a divinely mandated hierarchy:

> To proceed to that which is next in order, from God to spirits; we find, as far as credit is to be given to the celestial hierarchy of that supposed Dionysius the senator of Athens, the first place or degree is given to the angels of love, which are termed Seraphim; the second to the angels of light, which are termed Cherubim; and the third and so following places to ... angels of power and ministry; so as the angels of knowledge and illumination are placed before the angels of office and domination.

> To descend from spirits and intellectual forms to sensible and material forms; we read the first form that was created was light, which has a relation and correspondence in nature and corporal things, to knowledge in spirits and incorporeal things.[22]

---

[19]   Karlfried Froehlich, "Pseudo Dionysius and the Reformation of the Sixteenth Century," in Pseudo-Dionysius, *The Complete Works*, pp. 33–46.

[20]   *Ibid.*, pp. 43–4.

[21]   See Mohamed, *In the Anteroom*.

[22]   *WFB*, III, p. 296.

Although Bacon introduces Dionysius in regard to the hierarchies of angels, he has been following the Areopagite's pattern of thought since introducing God as the originator of knowledge which emanates downward through the hierarchical system. Bacon continues to run in the tracks of Dionysius in his use of "light" as a metaphor for knowledge, which is prominent throughout the Dionysian corpus.

The proper path of learning is consistently described by Bacon as an ascent from the observation of particular instances to general principles, as in aphorism 19 of the *Novum Organum*, book 1:

> There are and can be only two ways of searching into and discovering truth. The one flies from the senses and particulars to the most general axioms, and from these principles, the truth of which it takes for settled and immoveable, proceeds to judgment and to the discovery of middle axioms. And this way is now in fashion. The other derives axioms from the senses and particulars, rising by a gradual and unbroken ascent [*ascendendo continenter at gradatim*], so that it arrives at the most general axioms last of all. This is the true way, but as yet untried.[23]

The "true" but "yet untried" way was an ascent which humans must achieve by negatives:

> To God, truly, the Giver and Architect of Forms, and it may be to the angels and the higher intelligences, it belongs to have an affirmative knowledge of forms immediately, and from the first contemplation. But this assuredly is more than man can do, to whom it is granted only to proceed at first by negatives, and at last to end in affirmatives, after exclusion has been exhausted.[24]

This was the path of true induction for Bacon [*Atque in exclusiva jacta sunt fundamenta Inductionis verae*].[25] It is entirely in keeping with the Dionysian hierarchy of angels that the "higher intelligences" should have an affirmative and immediate knowledge of creation, and that "from the first contemplation" [*ab initio contemplationis*]. For Pseudo-Dionysius the highest ranks of the angels are the creatures of the "first contemplation" because they do not need to know the workings of God via secondary effects, or "the symbols of the senses according to the intellect" [*non sicut sensiblilium symbolorum intellectualiter*].[26] This gives them a "supreme (and angelic) knowledge of the operations of God" [*supremam*

---

[23]   *WFB*, IV, p. 50. Latin, *WFB*, I, p. 159.

[24]   Aphorism 15, *Novum Organum* book 2, in *WFB*, IV, p. 145. Latin, *WFB*, I, pp. 256–7.

[25]   Aphorism 19, *Novum Organum* book 2, in *WFB*, I, p. 260.

[26]   *Celestial Hierarchies*, Chapter VII, in Pseudo-Dionysius, *The Complete Works*, p. 138. The Latin is Sarrazin's translation, widely available in Bacon's day and typical of Latin readings. See *Dionysiaca*, p. 847.

(ut in angelis) Dei operationum scientiam].[27] Humans, for their part, require sensible, created things, including scriptures, to obtain knowledge. As Pseudo-Dionysius puts it in the *Celestial Hierarchy*, "it is quite impossible that we humans should, in any immaterial way, rise up to imitate and to contemplate the heavenly hierarchies without the aid of those material means capable of guiding us as our nature requires."[28]

The major difference between the *apophatic* way of Pseudo-Dionysius and Bacon's ascent by negatives is not one of method, but subject. For Pseudo-Dionysius the subject of contemplation was always God. Bacon applied the same method to the study of nature. God may be studied both by affirmations (since certain truths about God have been revealed in the scriptures) and by the systematic negation of all that God is not. For Bacon, nature may not be studied by affirmations, since affirmatives have been hidden by God in nature itself, and are known only to God and the "higher intelligences" immediately. Bacon repeatedly illustrated this principle by referring to Proverbs 25:2, interpreted as sanctioning the human participation in divine knowledge: "it is the Glory of God to conceal a thing, but it is the glory of the King to find a thing out."[29] What Bacon proposed was the application of the negative way, the only way which can lead to certainty when one starts with the particulars of creation, to creation itself.

If Bacon's eliminative induction has a precedent in the Dionysian *via negativa*, it was still, when applied to natural philosophy, a new and "untried" way. As Bacon recounted the early centuries of the Christian era he observed that there was no progress in natural philosophy because "the greater number of the best wits applied themselves to Theology."[30] For Bacon, matters of the faith always came first. The instauration of knowledge of the natural world was possible in his day precisely because the final reformation of religion was already taking place:

> And we see before our eyes, that in the age of ourselves and our fathers, when it pleased God to call the church of Rome to account for their degenerate manners and ceremonies, and sundry doctrines obnoxious and framed to uphold the same abuses; at one and the same time it was ordained by the Divine Providence that there should attend withal a renovation and new spring of all other knowledges.[31]

Although Bacon never hesitated to criticize the Church of Rome, he was not a mainline Protestant in Tudor/Stuart England. He shared the conviction of his

---

[27]   *Ibid.*, pp. 850–51.

[28]   Pseudo-Dionysius, *The Complete Works*, p. 146.

[29]   See the preface to the *Instauratio Magna*, in *WFB*, IV, p. 20; Latin, *WFB*, I, p. 132. Compare *Valerius Terminus*, in *WFB*, III, p. 220, where Bacon explains that "King" means humankind.

[30]   *Novum Organum*, Spedding translation, in *WFB*, IV, p. 78.

[31]   *WFB*, III, p. 300.

friend Lancelot Andrewes, and a significant minority of like-minded intellectuals, that the Reformation required not an adoption of the commonplaces of Calvin, but a return to the theology of the patristic era, prior to the errors of Rome.[32] Along with this theological recovery, in the "last ages of the world," according to Bacon's famous interpretation of Daniel 12:4, the time would come for the advancement of all sciences:

> Nor should the prophecy of Daniel be forgotten, touching the last ages of the world:—"Many shall go to and fro, and knowledge shall be increased;" clearly intimating that the thorough passage of the world (which now by so many distant voyages seems to be accomplished or in course of accomplishment), and the advancement of the sciences, are destined by fate, that is by Divine Providence, to meet in the same age.[33]

Bacon believed that he was observing an age ordained by providence for the recovery of subjects for which the ancient Christian fathers did not have time: the recovery of the knowledge of the created order. The *via negativa*, part of classical theology for centuries, had not been applied to natural philosophy, but its time had come.

For Bacon, as he applied eliminative induction to the created order, the result would be positive answers which were higher in the hierarchy of knowledge than the particulars with which he started. In the process, the mind rose from particulars to general principles. Once these general principles could be intimated (as in heat being a variety of motion), the intellect could descend to particular instances again and test the hypothesis. In this way there was also a cataphatic element of Bacon's method. The Dionysian author wrote of the two ways of knowing: "When we made assertions we began with the first things, moved down through intermediate terms until we reached the last things. But now as we climb from the last things up to the most primary we deny all things."[34] Descent from principles was only possible if principles were known. For Pseudo-Dionysius the first principles of understanding divinity were clear from scripture. In Bacon's reasoning the principles of creation could be understood through reading the book of nature, using the ascent from particulars to principles by means of negatives, and then testing the conclusions to determine whether one was dealing with a true law of nature. The old way of reasoning in natural philosophy, by which past generations had flown "from the senses and particulars to the most general axioms," lacked the certainty which only negative induction could provide. In the same way, when it came to theological questions, the ultimate truth of the essence of God could only properly be approached by negations.

---

[32] See Matthews, *Theology and Science*, pp. 27–50.

[33] *WFB*, IV, pp. 91–2. Latin, *WFB*, I, p. 200.

[34] Pseudo-Dionysius, *The Complete Works*, p. 138.

In the *Mystical Theology*, the Dionysian author gave a model of an ascent by negation from created things to the state of unknowing in which the intellect can mystically contemplate God. Chapters four and five are what Bacon would call "Tables of Presentation":

> The Cause of all is above all [the Dionysian premise] and is not inexistent, lifeless, speechless, mindless. It is not a material body, and hence has neither shape nor form, quality, quantity, or weight. It is not in any place and can neither be seen nor touched … It passes through no change, decay, division, loss, no ebb and flow, nothing of which the senses may be aware.[35]

The difference in the subject of contemplation (God or creation) must necessarily lead to different results based upon a fundamental distinction between the subjects. While the things of creation are finite and may be defined according to their essential natures, God transcends everything. When Bacon applied the process of negation to natural phenomena, he eventually arrived at positive answers. When the *via negativa* is applied to a God who is both immanent in creation and yet transcends all things (including concepts) the "answer" is quite different. For Pseudo-Dionysius the application of the denials to the subject of God was the paradox of unknowing: "We deny all things so that we may unhiddenly know that unknowing which itself is hidden from all those possessed of knowing amid all beings, so that we may see above being that darkness concealed from all the light among beings."[36]

Knowledge and Power

Bacon shared the Dionysian conclusion of the unknowability of God. According to Bacon, knowledge of creation could not ultimately reveal anything of the nature or will of God:

> For if any man shall think by view and inquiry into these sensible and material things, to attain to any light for the revealing of the nature and will of God, he shall dangerously abuse himself. It is true that the contemplation of the creatures of God hath for an end (as to the natures of the creatures themselves) knowledge, but as to the nature of God, no knowledge, but wonder: which is nothing else but contemplation broken off, or losing itself.[37]

In Bacon's system, natural philosophy produced, through a consistent application of negations, the conclusion of the hiddenness of God. Physical images could be

---

[35]   *Ibid.*, pp. 140–41.

[36]   *Ibid.*, p. 138.

[37]   *WFB*, III, p. 218.

used to represent certain truths about God, according to the *Celestial Hierarchies*, but "God is in no way like the things that have being and we have no knowledge at all of his incomprehensible and ineffable transcendence and invisibility."[38] This is the same concept expressed by Bacon when he avers that "there is no proceeding in invention of knowledge but by similitude; and God is only self-like, having nothing in common with any creature otherwise than as in shadow and trope."[39]

Any encounter with the true nature of God cannot occur in the realm of things that are intelligible, since that would only be an encounter with a creaturely representation of God. As the Dionysian author expressed it in the *Divine Names*, "the union of divinized minds with the Light beyond all deity occurs in the cessation of all intelligent activity."[40] This "cessation" is the same point of mystic ascent identified by Bacon when the contemplation of the creatures leads to "wonder; which is nothing else but contemplation broken off or losing itself."[41]

There could be, for both Bacon and Dionysius, no knowledge of God's transcendent nature, but both acknowledged that through theology God drew humankind upward to truths beyond rational discovery, even as reason strove after divine truth. As Bacon explained the "mysteries of God":

> *Da fidei quae fidei sunt*: [give unto Faith that which is Faith's]. For the Heathen themselves conclude as much in that excellent and divine fable of the golden chain: *That men and gods were not able to draw Jupiter down to the earth; but contrariwise, Jupiter was able to draw them up to heaven.* So as we ought not to attempt to draw down or submit the mysteries of God to our reason; but contrariwise to raise and advance our reason to the divine truth.[42]

God is both the origin of knowledge and the goal of reason for Bacon, and this reflects the writings of Dionysius: "The divine Wisdom is the source, the cause, the substance, the perfection, the protector, and the goal of Wisdom itself, of mind, of reason, and of all sense perception."[43] Where human reason cannot go by itself, toward understanding the "mysteries of the faith," God can draw it. Pseudo-

---

38   Pseudo-Dionysius, *The Complete Works*, p. 150.

39   *WFB*, III, p. 218.

40   Pseudo-Dionysius, *The Complete Works*, p. 54. The force of this sentence on "cessation" (*cessationem*) rather than "resting from" or "quieting" (*requietionem*) reflects later Latin translators. See *Dionysiaca*, p. 38.

41   *WFB*, III, p. 218.

42   *Ibid.*, III, p. 350. Italics original. This is the original published appearance of the maxim, *Da fidei quae fidei sunt*, found later in Aphorism 65 of the *Novum Organum*; see *WFB*, I, p. 245. Consider the similar image in *The Advancement of Learning, ibid.*, III, p. 268. The hierarchy of knowledge above nature belongs to revealed theology. The hierarchy of knowledge of nature is the province of the study of natural philosophy.

43   Pseudo-Dionysius, *The Complete Works*, p. 107.

Dionysius made the same point, significantly using the same image from book eight of Homer's *Iliad*:

> Imagine a great shining chain hanging downward from the heights of heaven to the world below. We grab hold of it with one hand and then another, and we seem to be pulling it down toward us. Actually it is already there on the heights and down below and instead of pulling it to us we are being lifted upward to that brilliance above … We will not pull down to ourselves that power which is both everywhere and yet nowhere, but by divine reminders and invocations we may commend ourselves to it and be joined to it.[44]

In all knowledge, hierarchy is the governing principle. At the lowest levels are the physical things of the created order (what we now call "nature") which were designed to be grasped by the human mind through simple reason. Above these are the mysteries which cannot be ascertained by human reason alone, but may still be apprehended by the intellect because they have been revealed through the Faith. Finally, the transcendent God exists in wonder beyond knowing.

For Bacon, knowledge of the created order extended beyond what we would recognize as "science" today, and it blurred what we regard as the boundary between the "natural" and the "supernatural." In *Valerius Terminus* Bacon stated: "whatever is not God but parcel of the world, he hath fitted it to the comprehension of man's mind, if man will open and dilate the powers of his understanding as he may."[45] Everything below the transcendent God, the entire order of created things, was potentially knowable, and was part of Bacon's restoration of the sciences. This included a Dionysian study of angels:

> Otherwise it is of the nature of angels and spirits, which is an appendix of theology both divine and natural, and is neither inscrutable nor interdicted; for although the Scripture saith, *Let no man deceive you in sublime discourse touching the worship of angels, pressing into that he knoweth not*, &c. yet notwithstanding if you observe well that precept, it may appear thereby that there be two things only forbidden, adoration of them, and opinion fantastical of them; either to extol them further than appertaineth to the degree of a creature, or to extol a man's knowledge of them further than he hath ground. But the sober and grounded inquiry which may arise out of the passages of holy Scriptures, or out of the gradations of nature, is not restrained.[46]

The study of angels or heavenly powers depended upon the use of the legitimate sources of scripture and the hierarchy of nature. Nothing was more in keeping with the approach of the Areopagite in the *Celestial Hierarchies*, where the scriptures

---

[44]   *Ibid.*, pp. 68–9.
[45]   *WFB*, III, p. 221.
[46]   *Ibid.*, p. 350. Compare *De Augmentis*, in *WFB*, I, p. 546.

were combined with the analogy of nature so that the celestial hierarchies could be understood.[47]

In both the Dionysian corpus and Bacon's work, "knowledge" (*scientia*) figures prominently as the means of ascent and descent through the created order. On account of this, creation may be seen in the writings of both as what Eric Perl aptly called the "continuum of cognition."[48] The highest order of angels is described by Dionysius as an "outpouring of wisdom" (*effusio sapientiae*) which passes the divine wisdom down the hierarchy to the next order, characterized by their reception of the divine light, at which point the *sapientia*, or divine wisdom of those who minister to God immediately, becomes the *scientia*, or knowledge, of those who receive the light mediately, as it passes downward from the second order.[49] The common distinction made in Latin versions of the *Celestial Hierarchies* between *scientia* and *sapientia* is also made by Bacon in *The Advancement of Learning*: "for all learning is knowledge acquired, and all knowledge in God is original: and therefore we must look for it by another name, that of wisdom, or sapience, as the Scriptures call it."[50] As the Dionysian author leads his readers down the hierarchies toward material creation it becomes clear that "perfecting knowledge" [*perfectivae scientiae*] is not mere awareness of truth or facts, as we tend to understand "knowledge"; it is a transforming power.[51]

It is often forgotten that Francis Bacon's most quoted axiom, "knowledge is power," was formulated in the context of a theological discussion. In his essay on "heresy" in the *Meditationes Sacrae*, Bacon wrestled with the concepts of "free will" and sin. On the one hand, he rejected the assertion that God was necessarily the author of evil, "not because he is not author—but because not of evil."[52] On the other hand, he rejected an opposing view which suggested God's power was limited: the position of those "who give a wider range to the knowledge of God than to his power; or rather to that part of God's power (for knowledge itself is power) [*nam et ipsa scientia potestas est*] whereby he knows, than to that whereby he works and acts; suffering him to foreknow some things as an unconcerned looker on, which he does not predestine and preordain."[53] Bacon's solution to these opposite errors was via an Augustinian chain of causes whereby sin could

---

[47]   *Dionysiaca*, p. 730. On material hierarchies, see p. 7.

[48]   Eric D. Perl, Theophany: *The Neoplatonic Philosophy of Dionysius the Areopagite* (Albany, 2007), pp. 83–100.

[49]   *Dionysiaca*, pp. 837–43. This distinction follows the original Greek: *sapientia* is usually a translation of the Greek *sophia*, originating in God, and *scientia* often translates a word based on *pistis* implying receptivity.

[50]   *WFB*, III, p. 295.

[51]   *Dionysiaca*, p. 843. The next rank of angels receives this knowledge and manifests it as power. Note Bacon's observation of this in his reference to the *Celestial Hierarchies* above.

[52]   *WFB*, VII, p. 254.

[53]   *Ibid.*, p. 253. Latin, p. 241.

be moved down the chain into human will, and yet be accounted for by God in his foreknowledge.[54] God would remain author of all as the ultimate cause, but he was author by "links and subordinate degrees" [*per nexus et gradus subordinatos*].[55] The chain of causes was crucial for Bacon's scientific thought. Natural philosophy was the study of these secondary causes according to Bacon, and "certain it is that God worketh nothing in nature but by second causes."[56]

For Dionysius, as for Bacon and Augustine, God always dealt with creation mediately. This was the reason for the cosmic hierarchies. For Dionysius, as for Bacon and Augustine, God was the author of all, but not of evil, since evil was not a created thing, but a defect, or an "accident."[57] But Pseudo-Dionysius provides the pivotal idea that in God knowledge and power were the same. It is crucial to recognize that Bacon never made the claim "knowledge is power" as a general axiom, but specifically in reference to God. For God, "knowledge itself is power," but *for humans*, knowledge and power could be very different. Indeed, in the very passage where Bacon directly references Dionysius, he does so in order to demonstrate that in human actions knowledge must *precede* power, as it does in the *Celestial Hierarchies*.

In the *Divine Names* Pseudo-Dionysius discusses God's action of creation in which knowing causes being: "He knows everything else and, if I may put it so, he knows them from the very beginning and therefore brings them into being."[58] Knowledge of creation and producing it must be the same in God, since God, for the Dionysian author, was a true Platonic monad, a singularity (an undivided Trinity) in which there could be no distinctions according to essence. The act of creation was a single emanation or outward expression of the Godhead itself.[59] According to Pseudo-Dionysius:

> Uniquely [the Divine Wisdom] knows and produces all things by its oneness: material things immaterially, divisible things indivisibly, plurality in a single act. If with one causal gesture God bestows being on everything, in that one same act of causation he will know everything through derivation from him and through their preexistence in him, and, therefore, his knowledge of things will not be owed to the things themselves.[60]

---

[54]   Compare Augustine, *De Civitate Dei*, V, 8–9. See also Matthews, *Theology and Science*, pp. 32–8.

[55]   *WFB*, VII, p. 253. Latin, p. 242.

[56]   *Ibid.*, III, p. 267.

[57]   See the discussion of evil in Pseudo-Dionysius, *The Complete Works*, pp. 93–6.

[58]   *Ibid.*, p. 107; closer to the Greek than the Latin translations. See *Dionysiaca*, pp. 396–7.

[59]   Perl, *Theophany*, pp. 17–34.

[60]   Pseudo-Dionysius, *The Complete Works*, p. 108.

Knowledge and power may be distinguished only in the differentiated hierarchies of creation, not in the undifferentiated Creator, or in his indivisible act of creation. In the *Divine Names*, knowledge and power are discussed as separate attributes of God, but only because they are human ways of identifying and describing the effects of the single divine act of emanation which was the creation of the cosmos.[61]

Pseudo-Dionysius opened the *Celestial Hierarchies* with James 1:17: "Every good endowment and every perfect gift is from above, coming down from the Father of Lights."[62] The rest of the discourse examines how the single ray proceeding from God produces all the diversity and differentiation found in the created hierarchy.[63] From this diversity the intellect could ascend back to the unity which is the source of all. Bacon's apophatic approach to discovery operated within the framework of this hierarchy. To obtain the knowledge of natural things, Bacon's method stopped part way, pondering the natural order itself, rather than the mystery of the transcendent God, though it could go further and break off in "wonder." Knowledge, being for Bacon "original in God," was the key to recovering power over creation, but only if one followed the proper path to obtain it. This *via negativa* of creation was Bacon's hermeneutic of discovery, his sure method, for recovering the power over nature which had been lost to Adam in Eden. As Bacon closed the *Novum Organum*: "For man by the fall fell at the same time from his state of innocency and from his dominion over creation. Both of these losses however can even in this life be in some part repaired; the former by religion and faith, and the latter by arts and sciences."[64]

Antonio Perez-Ramos has identified Bacon's method with the "maker's knowledge tradition" whereby knowing and making are inextricably linked.[65] The Dionysian understanding of God's bringing forth the cosmos in a single act of creating/knowing provides a source for this tradition in Bacon's writings. Bacon's method, by uniting experimentation with the divine gift of knowledge, would raise humankind to being what it was created to be: the very image of the making/knowing God. The key was recognizing how knowledge worked within the divine hierarchy of creation: it originated in God, and through the *via negativa* it would return to Him as humans would "raise and advance [their] reason to the divine truth."[66] Along the way, this ascent would restore Adam's "dominion over creation." The result was guaranteed by none other than the "Father of Lights." Citing the same verse with which the Dionysian author began, Bacon concluded: "The beginning is from God: for the business which is in hand, having the character

---

[61]   Compare chapters seven and eight of the *Divine Names*.

[62]   Pseudo-Dionysius, *The Complete Works*, p. 145.

[63]   *Ibid.*, p. 146.

[64]   *WFB*, IV, p. 247.

[65]   Antonio Perez-Ramos, *Francis Bacon's Idea of Science and the Maker's Knowledge Tradition* (Oxford, 1988), pp. 54–62.

[66]   *WFB*, III, p. 350.

of good so strongly impressed upon it, appears manifestly to proceed from God, who is the author of good and the Father of Lights. Now in divine operations even the smallest beginnings lead of a certainty to their end."[67]

Conclusion

There can be no question that Pseudo-Dionysius had an influence on Bacon's conception of the hierarchies of knowledge and the hierarchical nature of creation, since Bacon himself references the Areopagite in precisely this context. In the Protestant environment of Stuart England, stating his affinity for the thought of Pseudo-Dionysius was not a minor thing. This chapter has argued that, when taken on Bacon's own terms, the central, and otherwise puzzling, adherence of Bacon to what we have called "eliminative induction" is a result of an agreement with Dionysius on the fundamental nature of "knowledge," its origin (in God), and its ultimate end (contemplation of the divine).

For students of Bacon, this argument leads in a direction directly counter to his more common narrative role as the instigator of a non-theological, and non-religious, study of nature. This old narrative requires us to ignore the extensive theological discussions found throughout Bacon's writings. It appears to be an atavistic remnant of the old "Age of Discovery" narrative which, as observed in this volume's Introduction, has met its undoing, or deconstruction, elsewhere. To be sure, Bacon may have had a personal affinity for Neoplatonism, but in Bacon's day, Neoplatonism could not be separated from its distinctly Christian profile both in antiquity and the Renaissance. A careful comparison makes it clear that Bacon did not need to do any tinkering to adapt Neoplatonism to a Christian natural philosophy. He followed to its logical conclusion the way prepared by Pseudo-Dionysius, somewhat incredulously commenting along the way that he was surprised that this remained untried. As with the ancient author, Bacon was concerned, first and last, with knowledge, and how it functioned in a divinely-structured cosmology. Before the ascent of atheism in the Enlightenment, knowledge could not be separated from its author, for whom, uniquely, knowledge (*scientia*) was power (*potestas*).

---

[67]   *WFB*, IV, p. 91.

Chapter 3

"Invention" and "Discovery" as Modes of Conceptual Integration: The Case of Thomas Harriot

Michael Booth

"Inventing" something, nowadays, means causing it to be; "discovering" something means perceiving that it already is. The logical opposition that currently pairs these terms, however, seems not to have done so consistently in early-modern usage. There, "discover" primarily meant uncover or reveal—as often a matter of showing to a second party as perceiving for oneself. "Invent," meanwhile, retained its Latin sense of *find*. If we go back far enough in early-modern English, we arrive at a point where people could speak colloquially of the pleasure of inventing (coming upon) what God or nature had discovered (shown) to them, whereas we might now idiomatically say almost the reverse. Yet we cannot necessarily infer that they simply or straightforwardly *meant* the reverse. Perhaps all we can really be sure of, with regard to early-modern "invention" and "discovery," is that both denote aspects of the experience of *novelty*; but with mutual and shifting imbrications of agency and subjectivity.

"Conceptual integration" or "blending," as explored by recent work in cognitive science, offers a useful way of understanding novelty in terms both of individual experience and of cultural or categorical change. If we sense that "making" and "finding" are interestingly overdetermined in early modernity, we may want to know about some recent efforts to describe these phenomena in cognitive terms.[1]

---

[1] For expositions of blend-theory, see Gilles Fauconnier and Mark Turner, *The Way We Think: Conceptual Blending and the Mind's Hidden Complexities* (New York, 2002), pp. 269–78; and Seana Coulson, *Semantic Leaps: Frame-Shifting and Conceptual Blending in Meaning Construction* (Cambridge, 2006). On blend-theory and Elizabethan culture, see Eve Sweetser, "'The suburbs of your good pleasure': Cognition, Culture and the Bases of Metaphoric Structure," in G. Bradshaw, T. Bishop, and M. Turner (eds), *The Shakespearean International Yearbook, vol. 4: Shakespeare Studies Today* (Aldershot, 2004), pp. 24–55; and Nicholas R. Moschovakis, "Topicality and Conceptual Blending: *Titus Andronicus* and the Case of William Hacket," *College Literature*, 33/1 (2006): pp. 127–50. For cognitive approaches to early-modern and literary studies generally, see F.E. Hart, "Matter, System and Early Modern Studies: Outlines for a Materialist Linguistics," *Configurations*, 6 (1998): pp. 311–43, and "The Epistemology of Cognitive Literary Study," *Philosophy and Literature*, 25 (2001): pp. 314–34.

The focus of the current collection is on our "understanding of understanding" as well as our understanding of the early-modern period, with the presumption that these two matters are intertwined and mutually illuminating. Recent contributions of cognitive linguists to the former issue, I believe, offer new opportunities for insight into the latter.

One early-modern person who was known for his involvement in both invention *and* discovery—however exactly we are to understand these terms—was the Elizabethan scientist and explorer Thomas Harriot.[2] "Imployed in discovering" is how Harriot characterizes his role as naturalist, ethnographer, and linguist for the 1585 Virginia expedition; "invention" is the word used of his mathematical and scientific work by his friend William Lower.[3] Harriot's name has been familiar to literary studies ever since New Historicism first emboldened scholars to read historical and literary texts into each other; he has gained a belated notoriety as an agent of colonialism, and has figured often in works of literary criticism and cultural history devoted to the subject.[4] What has found little place in these discussions, however, is serious consideration of Harriot's Algonquian-language experience—let alone its potential relationship to his mathematical innovations.[5]

---

[2]   See Stephen Clucas, "Thomas Harriot and the Field of Knowledge in the English Renaissance," in Robert Fox (ed.), *Thomas Harriot: An Elizabethan Man of Science* (Aldershot, 2000), pp. 93–136; and John W. Shirley, *Thomas Harriot: A Biography* (Oxford, 1983), and (ed.), *A Source Book for the Study of Thomas Harriot* (New York, 1981), and (ed.), *Thomas Harriot: Renaissance Scientist* (Oxford, 1974). On Harriot's linguistics, see Vivian Salmon, "Thomas Harriot and the Origins of Algonkian Linguistics," *Language and Society in Early Modern England* (Amsterdam, 1996), pp. 143–72.

[3]   Harriot, *A Briefe and True Report of the New Found Land of Virginia* (New York, 1972), title page; letter of Lower to Harriot, quoted in Shirley, *Thomas Harriot*, p. 400.

[4]   See Stephen Greenblatt, "Invisible Bullets," *Shakespearean Negotiations* (Berkeley, 1988), pp. 21–65; Eric Cheyfitz, *The Poetics of Imperialism: Translation and Colonization from "The Tempest" to "Tarzan"* (New York, 1991); Gesa MacKenthun, *Metaphors of Dispossession: American Beginnings and the Translation of Empire, 1492–1637* (Norman, 1997); Bruce R. Smith, "Mouthpieces: Native American Voices in *Thomas Harriot's True and Brief Report [sic] of ... Virginia*, Gaspar Pérez de Villagrá's *Historia de la Nueva México*, and John Smith's *General History of Virginia*," *New Literary History*, 32 (2001): pp. 501–17; Amir Alexander, "The Imperialist Space of Elizabethan Mathematics," *Studies in History and Philosophy of Science*, 26/4 (1995): pp. 559–91; and Deborah Jacobs, "Critical Imperialism and Renaissance Drama: The Case of 'The Roaring Girl'," in Dale M. Bauer and Susan Jaret McKinstry (eds), *Feminism, Bakhtin, and the Dialogic* (Albany, 1991), pp. 73–84.

[5]   For the latter, see Jacqueline Stedall, *The Greate Invention of Algebra: Thomas Harriot's "Treatise on Equations"* (Oxford, 2003); R.C.H. Tanner, "On the Role of Equality and Inequality in the History of Mathematics," *The British Journal for the History of Science*, 1/2 (1962): pp. 159–69; Helena M. Pycior, *Symbols, Impossible Numbers, and Geometric Entanglements* (Cambridge, 1997); and Edward Rosen, "Harriot's Science: The Intellectual Background," in Shirley (ed.), *Thomas Harriot: Renaissance Scientist*, pp. 1–15.

In what follows, I hope to use the theory of conceptual blending to make some sense of Harriot's multiple and seemingly disparate experiences of novelty. That is to say, I hope to show that they are really *one* experience, in a way that is theoretically, as well as historically, informative. I do not claim that Harriot's Algonquian translation work must have informed his mathematical thought—but I think I can offer a reasonable account of how it could have: how his invention of algebraic techniques might have flowed from his discovery of Algonquian syntactic relationships, and how his "discovery" of mathematical principles may reflect an inventive mind at work, knitting together the conceptual relationships of early-modern mathematics with those suggested by his encounter with an unfamiliar language and culture. In sum, I will argue here for Harriot's work as an exemplary, if exotic, blend of early-modern invention and discovery; and for the theory of conceptual blending as a useful, though relatively unfamiliar, way of talking about the latter.

Letters and Numbers

One of the most consequential of all historical developments in mathematics, it has been argued, was "the adoption, in the decades around 1600, of literal symbolism—that is, the use of letter symbols to represent arbitrary or unknown numbers."[6] Harriot was an advanced figure in this development; his work had greater symbolic abstraction than that of the early algebraic pioneer Viéte, and it preceded that of the more celebrated Descartes.[7] Jacqueline Stedall, the leading authority on Harriot's mathematical work, has written that "as early as 1600 he was already writing purely symbolic mathematics. With hindsight this can be seen as an early manifestation of a more general trend, but at the same time it can be recognized that it sprang from Harriot's innate interest and ability in such matters, the evidence for which begins with his Algonquian alphabet of 1585."[8] The latter, she says, "was not just a phonetic alphabet but something very like a phonetic algebra"[9]—a system in which observed relationships among phonemes are systematically correlated with observable relationships among the letter shapes that "encode information about the position and formation of each sound in

---

[6]   John Derbyshire, *Unknown Quantity: A Real and Imaginary History of Algebra* (Washington, DC, 2006), p. 3.

[7]   Florian Cajori, "A Revaluation of Harriot's *Artis Analyticae Praxis*," *Isis*, 11 (1928): pp. 316–24; pp. 317–18. Jacqueline Stedall says "Harriot's mathematics is almost wordless, because he expects (and he is almost always right) that his reader will be able to *see* what he is doing either by following a symbolic argument, or from the layout of his material on the page." Stedall, "Symbolism, Combinations, and Visual Imagery in the Mathematics of Thomas Harriot," *Historia Mathematica*, 34 (2007): pp. 380–401; p. 383.

[8]   Stedall, "Symbolism," p. 400.

[9]   *Ibid.*, p. 383.

the mouth."[10] The question for us, here, is what relationship may obtain between Harriot's mathematical innovations and his empirical encounter with Algonquian languages.

Regrettably, the record of Harriot's Algonquian experience is scant. Nonetheless, we do have some testimony about how the peculiarities of Algonquian grammar struck some other early-modern Englishmen. John Eliot (1604–90), while noting the frequently substantial length of individual Algonquian words on account of "the many syllables which the grammar rule requires," also says the language "doth greatly delight in the compounding of words ... to *speak much in few words*."[11] Roger Williams (1603–83) notes a grammatical form that he likens to the ablative absolute in Latin, and that enables the Narragansetts to "comprise much in little."[12] Peter Stephen Du Ponceau (1760–1844) writes that "the mind is lost in the contemplation of the multitude of ideas thus *expressed at once* by means of a single [Algonquian] word, varied through moods, tenses, persons, affirmation, negation, transitions, etc., by regular forms and cadences."[13] John Pickering (1777–1846) notes "a new manner of compounding words from various roots, so as to strike the mind at once with a whole mass of ideas."[14]

Some categories in Algonquian grammar seem particularly pertinent to Harriot's mathematical work. There is, for example, what Eliot calls the "suffix animate *mutual*"—relevant because of the intrinsic reciprocity of the relations among elements in algebra—and "the suffix form advocate, or '*instead*' form, when one acteth in the room or stead of another."[15] There is also a "distributive" inflection: distributives "express the number of things taken at a time, as 'each one, two at a time, every third one, four apiece',," which seems similar to what algebraic variables and coefficients do. William Jones cites its use in multiplication: "*ne'sw'*, three; ... *ne'swi nä'nese'nw'*, three is taken thrice at a time."[16] Distributives are indicated syntactically "by means of reduplication."[17] As John Eliot says, "when the action is doubled, or frequented, &c., this notion ... is expressed by doubling the first syllable of the word: as *Sa-sabbathdayeu*, every Sabbath."[18] This

---

[10]   *Ibid.*

[11]   John Eliot, *A Grammar of the Massachusetts Indian Language by John Eliot. A New Edition: With Notes and Observations by Peter S. Du Ponceau, L.L.D. and an Introduction and Supplementary Observations by John Pickering* (Boston, 1822), p. 6.

[12]   Roger Williams, *A Key into the Language of America* (1643) (Detroit, 1973), p. 56.

[13]   "Notes on Eliot's Indian Grammar," in Eliot, *A Grammar*, p. xxii.

[14]   "The Massachusetts Language: Introductory Observations," in Eliot, *A Grammar*, p. 4.

[15]   Eliot, *A Grammar*, p. 17.

[16]   William Jones, "Algonquian (Fox)," in Franz Boas (ed.), *Handbook of American Indian Languages, Part 1, Volume 2* (Bristol, 2002), pp. 735–874; p. 865. Jones was himself a Native American of the Fox (Algonquian) nation.

[17]   *Ibid.*, p. 762.

[18]   Eliot, *A Grammar*, p. 17.

inflectional category, or categorizing inflection, is interesting in light of Harriot's algebra, where the multiplication of a quantity by itself any number of times is not indicated in the Cartesian way, with superscript numerical exponents (e.g. "a^4"), but by reduplication (e.g. "*aaaa*"). The pattern of quantitative distribution can extend to any number of "dimensions" whatsoever, and is not limited to the three that can be illustrated by length, area, and volume. "Harriot was the first to devise a notation in which any number of unknown quantities could appear to any power."[19]

The imaginative and conceptual extension necessary to distribute specified shares or reflections of hypothetical quantity *a* from the second- and third- to a "fourth dimension"[20] may to some extent correlate with what would be needed to extend one's grammar, in conversation, to accommodate the unaccustomed consideration of the Algonquian "fourth person." This is the "obviative" inflection, which obliges a speaker to observe and indicate the "relative contributions of two third persons to action within a sentence or utterance."[21] Leonard Bloomfield describes it in terms of "identity and non-identity of objects,"[22] which of course is a primary concern of algebra. Algonquian grammar indicates whether a given verb implicates the listener or the speaker—in that order, which in itself is noteworthy—or a "proximate" third party, or an "obviative" *other* party, or any combination of these. The constant reassigning of grammatical markers for the latter two categories, necessary as one's discourse plays over a wide field of subjects and entities, is called by linguists "proximate shifting."

Laura Ann Buszard, who studies the phenomenon in Potawatomi Algonquian discourse, explains that proximate shifts occur "when there is a focus on a

---

[19]   Stedall, "Symbolism," pp. 398–9.

[20]   "Among the papers at Petworth, Rigaud identified a complete resolution by Harriot of a quartic equation with one positive, one negative and two complex roots." Gordon R. Batho, "Thomas Harriot's Manuscripts," in Robert Fox (ed.), *Thomas Harriot: An Elizabethan Man of Science* (Aldershot, 2000), pp. 286–97; p. 291. Harriot "was the first mathematician ever to solve a biquadratic completely for all four roots, positive, negative and imaginary." Jacqueline Stedall, "Rob'd of Glories: The Posthumous Misfortunes of Thomas Harriot and His Algebra," *Archive for History of Exact Sciences*, 54 (2000): pp. 455–97; p. 480.

[21]   Kathleen Bragdon, "Native Languages as Spoken and Written: Views from Southern New England," in Edward G. Gray and Norman Fiering (eds), *The Language Encounter in the Americas, 1492–1800* (Oxford, 2000), pp. 173–88; p. 178. Bragdon notes that this inflection was known to John Eliot.

[22]   "The Algonquian languages use different forms for non-identical animate third persons in a context. In Cree, if we speak of a man and then, secondarily, of another man, we mention the first one as ['na:pe:w] 'man', and the second one, in the so-called *obviative* form, as ['na:pe:wa]." Leonard Bloomfield, *Language* (New York, 1958), p. 193. The consideration of another entity, another "factor" in the speech-context, is here indicated by the addition of a single phoneme, *a*, which again interestingly prefigures Harriot's algebraic notation.

particular character, or the narrative is presented from a particular character's viewpoint," and that the effect is "to shift our attention or focus to a secondary character, or to represent that character's point of view."[23] Harriot records at least one Algonquian-language conversation with this kind and degree of complexity. "It was told me for strange news," he writes,

> that one being dead, buried and taken up again ... showed that although his body had lain dead in the grave, yet his soule was alive, and had travelled far in a long broad way, on both sides whereof grew most delicate and pleasant trees, bearing more rare and excellent fruits then ever he had seen before or was able to express, and at length came to most brave and fair houses, near which he met his father, that had been dead before, who gave him great charge to go back again and show his friends what good they were to do to enioy the pleasures of that place, which when he had done he should after come again.[24]

In her account of this sort of phenomenon as it occurs, with varying effects, in different Algonquian languages, Buszard relies on the theory of mental spaces: "the use of these grammatical constructions reflects a difference in the mental space configurations for foreground and background."[25] Indexing the participants in a discourse situation, providing "a conceptualization of viewpoint structure,"[26] is in fact the original purpose of mental space theory in linguistic analysis; as Ilana Mushin says, it is "designed to be able to track propositional information through different belief spaces and so represent how information is distributed through different viewpoints."[27] It is notable that Adrian Robert also employs the mental-space model in his discussion of how we track "the specific elements of an imagined configurational situation" in mathematical proofs.[28]

Harriot would have had to reckon with the way that Algonquian grammar forces a speaker to mark at every moment the array of viewpoints involved in a speech-situation—and that is just taking into account the human ones; Algonquian languages inflectionally distinguish animate from inanimate entities, but in a given context almost anything may become grammatically animate.[29] The cultural

---

[23]   Laura Ann Buszard, "Constructional Polysemy and Mental Spaces in Potawatomi Discourse" (Dissertation, Dept. of Linguistics, University of California, Berkeley, 2003), pp. 190, 183.

[24]   Harriot, *A Briefe and True Report*, p. 26.

[25]   Buszard, *Constructional*, p. 121.

[26]   Ilana Mushin, "Viewpoint Shifts in Narrative," in Jean-Pierre Koenig (ed.), *Discourse and Cognition: Bridging the Gap* (Stanford, 1998), pp. 323–36; p. 335.

[27]   *Ibid.*, p. 327.

[28]   See Adrian Robert, "Blending and Other Conceptual Operations in the Interpretation of Mathematical Proofs," in Koenig, *Discourse and Cognition*, pp. 337–50; p. 338.

[29]   "The animate gender is, as it were, *absorptive*. If an inanimate noun stands in apposition with a local pronoun [i.e. 'you' or 'I'], or if that named by the noun takes on

historian Howard S. Russell observes that, as "every man, bird, beast, flower, fruit, or even rock had its role and special value," a traditional Algonquian man "felt himself in the presence of living entities who were as conscious of his existence as he of theirs."[30] The discursive world into which Harriot entered, in venturing to converse with the native people of the Carolina Outer Banks, would have presented him repeatedly with the challenge of organizing semantic relationships among the entities participating or implicated in the discourse, variably foregrounding their presence or absence, and reshuffling the mental space configurations necessary to do so. Such juggling of spaces, I think, suggestively prefigures the algebra that Harriot pioneered.

Blending-Compression

Mental spaces, on the view of cognitive linguists, behave in two important ways: they link together to form networks which are salient or active for a period of time (as when one mentally co-ordinates variables in a mathematical problem, or persons in a narrative) and they also at times integrate or blend, as when a relationship of identity—whether exact and literal, or approximate and figurative—is found between them. Like poststructuralist paradigms, blend-theory is a constructivist model of meaning. In place of the deconstruction of *terms*, though, as a conception of what happens when perceptive analytical thinkers scrutinize or escape reductive dichotomies, blend-theory offers the unpacking or decompression of mental *scenarios*, or "frames," that have been combined and compressed to form any instance of meaning under examination.

The process of conceptual blending tends toward the condensing or compression of what is diffuse into an integrated, more tractable form. Concision in writing is an example of this compression: when we say that a student has boiled down her first draft, or has synthesized the readings she was assigned, we are talking about these phenomena. Compression is bound by some constraints: for example, a preference that the things compressed should not be irreparably distorted, and that the new conceptual synthesis should lend itself readily to unpacking or expansion.[31] Harriot's mathematical innovations included one manifestation

---

the power of speaking or of being addressed (as in a story), or in any other way is assigned powers usually associated with people, animals and spirits, it tends to become animate in that context and to be so treated syntactically, sometimes even inflectionally." Charles F. Hockett, "What Algonquian is Really Like," *International Journal of American Linguistics*, 32/1 (1966): pp. 59–73; p. 62.

[30]   Howard S. Russell, *Indian New England before the Mayflower* (Hanover, 1980), p. 44.

[31]   See Fauconnier and Turner, *The Way*, p. 329: "Since the essence of a conceptual integration network is to project from many different … inputs into a single blended space, integration in that space is a considerable achievement, not something implicit in the inputs.

of such compression in his devising of "canonical equations"—a repertoire of equation forms assembled so that complex higher-order equations could readily be solved if manipulated into fitting one of these patterns.[32] These equations, along with Harriot's shift from verbal to symbolic algebra, allowed far greater reaches of mathematical quantity to be choreographed into a smaller space than had previously been possible.[33] Harriot's canonical equations might be seen as accomplishing within the constraints of mathematical exactness something akin to the linguistic phenomenon of "chunking" complex predications into pronouns or into concise new nouns.[34]

Another principle of blending is that notions of similarity often reflect a perception of shared spatial topology between the things in question; we tend to think in terms of scenarios derived from our embodied experience, so things seem significantly similar to us when they share a quality that impinges on that experience. What's true of thought in general is also true of mathematical thought in particular, as it attends to the growing, shrinking, disappearance, displacement etc. of hypothetical quantities. With respect to literary and cultural studies, the attention that blend-theory pays to scalar or gradient relationships as a factor in meaning is one reason why it better accommodates the articulation of some insights than a critical discourse concerned with dichotomies would.[35] With respect to Harriot, I have wondered if a symbolic notation he devised for indicating linear, planar, or cubic quantity might not have been suggested by the Algonquian dimensional classifiers: grammatical inflections that indicate whether an item is elongated, flat,

---

Integration in the blended space allows its manipulation as a unit, makes it more memorable, and enables the thinker to run the blend without constant reference to the other spaces in the network. Integration helps bring the blend to human scale, and thereby also increase the possibility for further useful recruitments to the blend from the range of our knowledge that is already at that human scale."

[32] See Stedall, *The Greate Invention*, p. 467: "This was ... where Harriot began to develop his profound and far-reaching insight that polynomial equations could be built up, or composed, of linear or quadratic factors such as $(a - b)$ or $(aa - df)$ ... Step by step, starting with the quadratic $(a - b)(a + c) = 0$, Harriot set out to compile a list of 'canonical equations'. From the nature of their composition it was easy to see not only what the roots of each equation must be, but also the relationship of the roots to the various coefficients. Hence, any given equation could be compared with a 'canonical', and one would immediately have important information about the number and nature of its roots."

[33] See Stedall, "Rob'd of Glories."

[34] Recursive noun-phrase constructions are indispensably useful in ordinary communication, but quickly become unwieldy as they grow. Fauconnier and Turner, *The Way*, p. 386: "By contrast, there is the capacity of language to come up with a single word for the same content. This is called *chunking* ... [If] 'scad' comes to mean 'the scarf my aunt my uncle my father dislikes married bought,' then it becomes easy to put [it] through further recursions."

[35] See Ellen Spolsky, "Darwin and Derrida: Cognitive Literary Theory as a Species of Post-Structuralism," *Poetics Today*, 23/1 (2002): pp. 43–62.

or voluminous.[36] The salience of spatial relations in blend-theory provides new theoretical grounds for exploring this kind of hypothesis.

In the domain of human meaning, as opposed to that of mathematics, blending involves conceptual frames, and frames clash. Sometimes a conceptual network brings together two things that have the same frame.[37] Sometimes, though, it brings together mental spaces that have strongly clashing frames, and invites us to draw select relevancies from each frame, ignoring some others, in order to construct a new, blended conception. A blended conception that uses the frame from one input to organize elements from another is called a "single-scope" blend.[38] An example would be Sir Walter Raleigh's comment that minds retain the permanent trace of their early education just as leather drinking vessels retain the "savour" of their first liquor.[39] This metaphor uses a culturally familiar fact about leather drinking vessels to give force to an argument about education; but it does not use the idea of education to make an argument about drinking vessels. If it did—if Raleigh drank from a new mug and declared it to have been thereby "educated"—he would be running the blend backwards, and offering something like poetry. A blend that communicates inferences back *and* forth in this way is called "double-scope."

Harriot shows a remarkable capacity for double-scope blending. In his mathematical work it appears, for instance, as a farsighted flexibility about how the multiplication of binomials should determine the positive or negative valence of their terms—whence his mnemonic rhyme: "Yet lesse of lesse makes lesse *or* more; / Use which is best; keep both in store."[40] Gilles Fauconnier and Mark Turner, the exponents of blend-theory, write: "We stress that failing to satisfy a governing principle [such as that of preserving each frame of reference intact] does not necessarily mean that the resulting blend fails; on the contrary, constructing a useful double-scope blend often depends upon finding a suitable way to relax governing principles."[41] In this respect, there is a suggestive resonance between the Algonquian "absentative" inflection—which indicates that the thing referred to is absent or does not exist—and Harriot's innovation of the predicative or algebraic

---

[36]   See Stedall, "Symbolism," p. 388.

[37]   For example, a punk-folk band is still a *band*; it can easily have predicated of it all of the things that pertain to any band: performances, recordings, audiences, etc. The level at which the two mental spaces of "punk band" and "folk band" clash is, as it were, within the frame of a "band," below the frame level of organization; the two input conceptions propose conflicting notions about how loud the band will play its music and how the audience will dress and act, but not about whether there will be music or an audience.

[38]   See Fauconnier and Turner, *The Way*, pp. 126–31.

[39]   Sir Walter Raleigh, *The History of the World*, ed. C.A. Patrides (Philadelphia, 1971), pp. 45–6.

[40]   For full text of the poem, see Pycior, *Symbols*, p. 63.

[41]   Fauconnier and Turner, *The Way*, p. 340.

zero.[42] About the latter, Alfred North Whitehead has written that "it is not going too far to say that no part of modern mathematics can properly be understood without constant recurrence to it."[43] Like an Algonquian absentative, Harriot's zero paradoxically indicates a "this" that isn't here. I am not implying that Algonquian native speakers would have experienced the absentative inflection as paradoxical. But Harriot very likely would have. By that very token, it is possible that he learned from his Algonquian hosts how to be analytically comfortable with the enabling paradox of a zero that could anchor algebraic expressions and arbitrate among them.

Visible Bullets

One of the very few other seventeenth-century Europeans who left a record characterized by "insight and understanding of the Indians of the Southeast"[44] was John Lederer, a multilingual physician originally from Hamburg, whose three journeys of American exploration began and ended in the Algonquian-speaking zone known to Harriot, each of them beginning from a different one of the rivers flowing into the Chesapeake Bay (the York, the James, and the Rappahannock). Lederer travelled as far as the Appalachian mountains, encountering mainly Iroquian- and Siouan-speaking groups along the way, but there was contact between these groups and the coastal Algonquians, and there was at least some cultural continuity across the groups. Even "the remoter Indians," he records, customarily carried "their currant Coyn of small shells, which they call *Roanoack*."[45] That, of course, is also the name of the coastal island where Harriot lived in 1585. Impressed by their "reason and understanding,"[46] Lederer observed that the people among whom he travelled were able to "supply their want of letters" in three ways: with "Tales," with "Emblemes or Hieroglyphicks," and with "Counters."[47]

The "counters" mentioned here may be pertinent to Harriot's mathematical thought. Lederer says:

> For Counters, they use either pebbles, or short scantlings of straw or reeds
> … Their reeds and straws serve them in Religious Ceremonies: for they lay

---

[42]   See Ives Goddard, "Comparative Algonquian," in Lyle Campbell and Marianne Mithun (eds), *The Languages of Native America: Historical and Comparative Assessment* (Austin, 1979), pp. 70–132; p. 99.

[43]   Alfred North Whitehead, *An Introduction to Mathematics* (Oxford, 1948), p. 45.

[44]   William P. Cumming, *The Southeast in Early Maps*, 3rd edn (Chapel Hill, 1998), p. 79.

[45]   Lederer, *The discoveries of John Lederer in three several marches from Virginia to the west of Carolina and other parts of the continent* (London, 1672), p. 27.

[46]   *Ibid.*, p. 5.

[47]   *Ibid.*, p. 3.

them orderly in a Circle when they prepare for Devotion or Sacrifice, and that performed, the Circle remains still; for it is Sacreledge to disturb or to touch it: the disposition and sorting of the straws and reeds, shew what kind of Rites have there been celebrated, as Invocation, Sacrifice, Burial, etc.[48]

We have here the picture of a landscape decked with geometric forms carefully constructed to encode information in the spatial relationship among their elements, and in particular we have the picture of one recurring shape: a large circle sketched in small straight lines. This exemplifies Harriot's insight that "a continuum is an aggregate of tangencies."[49]

The "pebbles" mentioned by Lederer are also important, as giving rise to another geometric form in the landscape. "Where a Battel has been fought," he writes, "or a colony seated, they raise a small Pyramid of these stones, consisting of the number slain or transplanted."[50] It is striking that, on this account of what sounds like memorial cairn-building, the *number* of stones is underscored as centrally important. It is also significant that the construction is referred to as a "pyramid," rather than, say, a pile. Since a pyramid must be built from a base of given dimensions, to make this shape with a given number of stones—the number of people commemorated—one would need to know the numerical ratio of side to base to full three-dimensional figure; that is, how many stones per side would give the base for a pyramid containing x stones. Lederer's testimony, if we take it seriously, suggests at least a practical tradition among the indigenous people for reckoning these ratios, a cultural feature that would have been highly convergent with known interests of Harriot's.

Scholars have observed, among his papers, some notes on the stacking of what he called "bullets," which in Harriot's time meant "a small ball," not necessarily considered as a missile.[51] "Concerning piling," the notes run, "there are two questions: one, the number of bulletes to be piled being geven with the forme of the ground plat, to know how many must be placed in every ranke, with how many rankes in the said ground plat. The second—a pile being made—to know the number of bulletes therein conteyned."[52] These bullets are generally presumed to have been the cannonballs on Raleigh's ships, but even assuming that they were,

---

[48] Lederer, quoted in Cumming, *The Southeast*, p. 79.

[49] See Alexander, "The Imperialist Space," p. 588.

[50] Lederer later notes, at one stop, "Near this Village we observed a Pyramid of stones piled up together, which their Priests told us, was the Number of an *Indian* Colony drawn out by Lot from a Neighbour-Countrey over-peopled, and led hither by one *Monack*, from whom they take the Name of *Monakin*." Lederer, *Discoveries*, p. 9.

[51] See *OED*: "1578 Lyte Dodoens I. viii. 15 Upon the braunches ... there groweth small bullets or rounde balles. *Ibid.* IV. lv. 515 It [the Reed Grass] bringeth foorth his boullettes, or prickley knoppes in August."

[52] Harriot, quoted in Shirley (ed.), *Thomas Harriot: Renaissance Scientist*, p. 242. Shirley observes: "the chart is ingeniously arranged so that it is possible to read directly the

this need not have been the first or only context in which Harriot considered the question, either practically or theoretically. Harriot's report that the Algonquians understood infectious diseases to be the result of "invisible bullets" cast by the English has been widely noted. Here, by contrast, we have a suggestion that Harriot could have learned from the Algonquians how to make numbers visible through the stacking of bullets.

This extremely simple point connects very directly with some of Harriot's most advanced and abstract mathematical work. He was interested, not just in triangles and pyramids, but in triangular and pyramidal numbers, the ratios or "constant differences" between the number of units in edges, faces, and solid figures. This is what Harriot called his "doctrine of triangular numbers" or "*Magisteria magna*."[53] It is an approach to constant difference interpolation: the relationship of series and continua. According to Stedall,

> The key to Harriot's method, as his title suggests, is an understanding of the properties of what he called 'triangular numbers', and so we begin by looking at what Harriot himself knew or discovered about such numbers ... The simplest triangular numbers are those that arise from stacking pebbles or dots in equilateral triangles ... If such triangles are imagined stacked vertically in decreasing order of size, they form triangular pyramids (or tetrahedra) and the number of pebbles or dots in successive pyramids are clearly sums of consecutive triangular numbers ... Harriot called these numbers 'pyramidal'.[54]

For example, the number of pebbles in triangles (tiers, counting from the top) of increasing size are 1, 3, 6, 10, 15, 21; if one counts the number of pebbles contained in the new pyramid formed each time a new base-tier is added, one has 1, 4, 10, 20, 35. As Stedall says, "the number of pebbles or dots in successive pyramids are clearly sums of consecutive triangular numbers." (That is, $1 + 3 = 4$, $1 + 3 + 6 = 10$, $1 + 3 + 6 + 10 = 20$, $1 + 3 + 6 + 10 + 15 = 35$, etc.) "Each triangle (or pyramid) is defined by the number of objects in a single side, 1, 2, 3, 4, 5 ... for which reason these numbers were sometimes called 'laterals' or 'sides' or 'roots'. These in turn are made up from simple units 1, 1, 1, 1." When arranged in rows, as they are in some of Harriot's manuscripts, this array of numbers "is now usually known as Pascal's triangle," though Blaise Pascal was born 63 years after Harriot. The numbers in each row give the binomial coefficients, as Harriot found in his algebraic work. Moreover, "the lateral numbers exhibit a constant difference (namely, 1), the triangular numbers a constant second difference, pyramidals a constant third difference, and so on. Harriot's theory was concerned with *any*

---

number of cannon balls on the ground or in the pyramidal pile with triangular, square, or oblong base." *Ibid.*

[53]   See Jacqueline Stedall and Janet Beery (eds), *Thomas Harriot's Doctrine of Triangular Numbers: The "Magisteria magna"* (Freiburg, 2009).

[54]   *Ibid.*, pp. 4–5.

sequence of numbers that has constant first, second, third, or higher differences." The calculation of "constant differences" is involved in the interpolation of data points between other data points, as Harriot undertook to do in his determination of meridional parts ("the adjustments needed at each degree of latitude in order to calculate an accurate position on a constant compass bearing")—a practical instance of the relationship between a geometric continuum and a mathematical series.[55]

Of course, Harriot did not need to learn about pyramidal piles of simple units from the Algonquians. But what is an invention, if not the recognition of the possibility of novelty in the familiar? Christopher Miller and George Hamell have written of an "identity of berries and beads" that is attested in many Algonquian languages[56]—a conceptual equivalence that greatly enhanced the value of European beads as trade goods, as they powerfully combined a supernatural value associated with berries and another value associated with materials like crystal and copper, which shared qualities of brightness and hardness with the beads.[57] As Miller and Hamell observe, "what were 'toys' to Verrazzano and 'trash' to John Smith were to the Woodland Indians powerful cultural metaphors that helped them to incorporate novel items and their bearers into their cognitive world"; it is generally by such a process, the same authors affirm, that "the unknown is made known."[58] Thus a hard round thing, for the Algonquians, was not just a pebble or a bullet, but a representation of the very unit of significance and complexity. And so it is for Harriot, as he imagines stacking up triangular numbers.

---

[55] *Ibid.*, p. 15.

[56] It is "reported for the Abnaki, Chippewa-Ojibwa, Montaignais-Naskapi, Menominee, Proto-Algonquian, and Cherokee." Christopher Miller and George R. Hamell, "A New Perspective on Indian-White Contact: Cultural Symbols and Colonial Trade," *Journal of American History*, 73 (1986): pp. 311–28; 322n. If the Indigenous Americans' seemingly metaphoric diction suggested that they were innocent of the cultural considerations that made wine different from "blood," biscuits from "wood," pens from "snowshoes," and beads from "berries," the Europeans had yet to learn the Algonquian logic of categorization that made berries spiritually significant or that associated the principles of "life, mind, and greatest being" with the lightness and transparency to be found in "shells and crystals as well as the hair of grandfathers." *Ibid.*, p. 324.

[57] "Since beads were linguistically and conceptually associated with berries, when shell, crystal and native copper were rendered into beads, these substances became metaphorical berries. Hence the items became exponentially more powerful." *Ibid.*, p. 323. It is interesting to recall in this connection that the English word "bead" meant "a prayer" before it meant a small ball or other item on a thread, and that the connection between beads and prayers was culturally active and explicit with some Europeans in the New World.

[58] *Ibid.*, pp. 328, 321.

Conclusion

One of the few scholars to have considered Harriot's mathematics in relation to any aspect of his New World experience is Amir Alexander. Emerging from Alexander's account as particular concerns for Harriot are the elusive notions of void in physics and nullity in mathematics, and their relations respectively to physical substance and geometric continuum. And emerging as a distinct and consequential innovation of Harriot's is his solution of the Mercator problem—how to chart a straight course on a curving Earth—by dividing the curve into small sections and positing a straight line in each, the result being an approximated curve "composed of an infinite number of straight segments, each bearing a fixed ratio to its predecessor."[59] All of these matters are cited by Alexander in support of a claim that Harriot's mathematics express or reflect a culturally specific but at the same time ubiquitous "narrative of Elizabethan expansionism," wherein "supposedly smooth surfaces were found to be lined with deep cleavages, leading to great wonders in the interior."[60]

I am, at one level, in agreement, and at another dissatisfied. I think that Alexander has pinpointed, with what he calls a "narrative," just the kind of basic scenario, grounded in embodied physicality, that cognitive theorists call a "conceptual frame" and treat—persuasively, in my opinion—as an unconscious organizing principle and basic building-block of thought. Alexander has noted the seeming recurrence of a particular conceptual frame, and has put the case, plausible in principle, that this frame exerted a powerful influence in Elizabethan culture and thought. It seems to me, though, that he has greatly overstated the cultural specificity of the conceptual frame that one could call "finding a passage" or "discerning something hidden", and also that his use of Harriot's mathematical ideas to confirm what he posits to be true of Harriot's whole society does little justice to the specificity of Alexander's own consideration of those mathematical ideas, or even, finally, to their importance as ideas.

In this chapter, I have tried to treat Harriot's work more flexibly and pragmatically. Blend-theory, it seems to me, paradigmatically allows such an approach. It is a nuanced account of culture and its creations that takes seriously both individual subjectivity and individual agency. Harriot lived and worked at the intersection of two historical beginnings: that of English America, and that of modernity in science and mathematics. He was a key innovator, an individual historically significant for the power of his thought; not just "power" in the sense now most common in the humanities, but in the sense of an extraordinary ability

---

[59]   Alexander, "The Imperialist Space," p. 584. This explanation summarizes Jon V. Pepper, "Harriot's Earlier Work on Mathematical Navigation: Theory and Practice," in Shirley (ed.), *Thomas Harriot: Renaissance Scientist*, pp. 54–90; also Pepper, "Harriot's Calculation of the Meridional Parts as Logarithmic Tangents," *Archive for the History of Exact Sciences*, 4 (1967–68): pp. 359–413.

[60]   Alexander, "The Imperialist Space," p. 579.

to hold many things in the mind, to integrate them where possible, and to represent those dynamics abstractly. There was much, in the cultural landscape of "Virginia" and its challenge of language-decipherment, that Harriot may or might have learned about meaning-construction and thought processes: about possibilities of calculation latent in shells, straws, and stones; about solving the riddle of an unknown expression; about the linking, sorting, substitution, and distribution of mental spaces; about the logic of another language, the logic of his own, and the possibility of thinking outside language altogether. As he blended old world and new, I have argued, so Harriot may have blended mathematics and linguistics, invention and discovery.

Chapter 4

The Undiscoverable Country: Occult Qualities, Scholasticism, and the End of Nescience

James Dougal Fleming

The doctrine of occult qualities, as articulated and defended in early-modern scholasticism, remains imperfectly understood.[1] The doctrine correlates cognitive with empirical failure. In other words, it posits that phenomena such as magnetism, which frustrate or elude causal sensing, are to that extent beyond the possibility of knowledge. Scholars have tended to conflate occult qualities with the Galenic theory of hidden sympathies and antipathies; with the modification of that theory by Neoplatonism and Paracelsianism; and with the predilection of *those* theories for astrology and magic.[2] I will argue here, however, that the scholastic doctrine of occult qualities, as strictly construed in the period, is disjunctive, not conjunctive, with its esoteric (mis)representations. I will further argue that the disjunct illuminates an epistemological incontinence in esotericism—and perhaps in the Newtonian settlement that follows the latter. What happened to occult qualities in the scientific revolution was not just a matter of semantic redefinition, but of hermeneutic reinvention. The writ of scientific discovery, rendered unlimited by

---

[1]   But see John Henry, *The Scientific Revolution and the Origins of Modern Science*, 3rd ed. (New York, 2008), pp. 60–63, and "Occult Qualities and the Experimental Philosophy: Active Principles in Pre-Newtonian Matter Theory," *History of Science*, 24 (1986): pp. 335–81; William Newman, *Atoms and Alchemy: Chymistry and the Experimental Origins of the Experimental Revolution* (Chicago, 2006), pp. 129–70; Keith Hutchison, "What Happened to Occult Qualities in the Scientific Revolution?" in Peter Dear (ed.), *The Scientific Enterprise in Early Modern Europe: Readings from* Isis (Chicago, 1997), pp. 86–106, and "Dormitive Virtues, Scholastic Qualities, and the New Philosophies," *History of Science*, 29 (1991): pp. 245–78; and Ron Millen, "The Manifestation of Occult Qualities in the Scientific Revolution," in Margaret J. Osler and Paul Lawrence Farber (eds), *Religion, Science and Worldview: Essays in Honor of Richard S. Westfall* (Cambridge, 1985), pp. 185–216.

[2]   See Henry, *Scientific*, 61–2; Hutchison, "What Happened," 89–94; Millen, "Manifestation," pp. 186–90; and Stephen Gaukroger, *The Emergence of a Scientific Culture: Science and the Shaping of Modernity, 1210–1685* (Oxford and New York, 2006), p. 255.

the esoteric retheorization of the occult, linked a posture of investigative ignorance to a postulate of human omniscience.

The Scholastic Doctrine

Empirical phenomena, in (some prominent versions of) early-modern scholasticism, present themselves through their "qualities" (or "properties"): the effects or characteristics that they impress upon our minds. Normally, a thing's qualities are "manifest": direct, consistent, and unmistakable. We perceive, not only such qualities, but also what it is to perceive them. As Aquinas puts it, they "have a clear origin, about which there arises no doubt."[3] Fire, for example, presents the manifest qualities of heat, light, burning, smoke-production, ash-production, etc. Our perception of these qualities does not appear illusory or strange; we do not have to ask ourselves what is happening, when we perceive them. The rest of the world, for that matter, seems to encounter them much as we do. Fire lights a room, as well as it does a face; warms soup, just as it does a pair of hands. What is more, fire performs much the same range of operations, in much the same way, whenever it occurs. Nobody has ever seen a fire that cooled things down, or suddenly produced fish, or turned bread into blood. Our perception of fire, in sum, appears reliable and intelligible. Knowledge of fire, accordingly (on this scholastic view), is possible.[4]

Not so with those phenomena (a restricted set) that present occult qualities. Although these are consistent, like manifest qualities, occult qualities are consistently elusive, indirect or strange. We do not perceive what is happening, when we perceive them. Indeed, it is unclear whether—or not—we can validly say that we perceive them at all. Examples include some purgative plants (notably rhubarb); fabulous creatures, such as the remora (a fish with the power to stop ships); and static electricity. But the *locus classicus* of the doctrine, as already mentioned, is magnetism. Sennert notes that "we see the attraction that arises from the magnet," but "we do not perceive" the quality through which it imparts motion into the iron.[5] The effect, the quality, of the lodestone is to move the iron. Yet a finger placed between the two objects feels no motion. Neither do we see, or otherwise perceive, anything passing between them. We perceive that there is a

---

[3]   St. Thomas Aquinas, "The Letter of Thomas Aquinas *De occultis operibus naturae ad quendam militem ultramontanum*," trans. J.B. McAllister, in *The Collected Works of St. Thomas Aquinas, Electronic edition* (Charlottesville, 1993), pp. 21–30; p. 21.

[4]   See Millen, "The Manifestation"; Newman, *Atoms*; and Aristotle, *De Anima* and *Categories*.

[5]   "*Attractionem, quae fit a Magnete, videmus; qualitatem autem, per quam ille ferri motus fit, non percipimus.*" Sennert, *De occultis qualitatibus*, book 2 of *Hypomnemata Physica* (Frankfurt am Main, 1636), p. 48. For Sennert as Aristotelian, see Newman, *Atoms*.

quality—by definition, something perceptible; but we do not perceive whatever quality is there. The magnet evades, even vitiates, perception. *A fortiori*, it escapes cognition. For the way to empirical knowledge (again, on this scholastic view) is supposed to start with a thing's quality. If the latter cannot even be grasped, then the construction of knowledge about the target phenomenon cannot even begin.

The usual way of putting this is to say that what is truly occult, in the world around us, cannot be made manifest. In early-modern scholasticism, this is partly for semantic, but primarily for epistemological, reasons. "It seems a kind of foolishness," writes the sixteenth-century natural philosopher Giovanni Olmo, "to seek the reason for a thing that we call occult, and actually perceive as such. Indeed, to call something occult, and then to seek the reason of it, implies a contradiction, and demonstrates tremendous ignorance."[6] The seventeenth-century Aristotelian Alexander Ross, polemicizing against the Copernican John Wilkins, finds the latter saying that *"the eye is an ill judge of naturall secrets."* "You should have said," Ross retorts, "that it is no judge of naturall secrets." For "the visible workes of nature are no secrets"; while "nature's secrets are invisible," by definition.[7] Ross's older contemporary Sennert makes clear that the issue here is not merely one of usage. The Aristotelian chemist inveighs against people who accept occult qualities, based on the observational evidence of magnetism etc.; but then dare to explain or illuminate the occult designation that they have supposedly granted, in terms of elemental mixtures.[8] It is precisely those who try to explain away the nature of occult qualities, for Sennert, who are playing around with the facts. Those, by contrast, who want to understand how things are, have to accept that occult qualities are *a priori* and *de re* unintelligible.

The occult, in sum, is a technical and local opacity within scholastic epistemology. As such, it is decidedly not a synonym or catch-all for the mysterious, or the unapparent. Scholastic defenders of occult qualities theorize the doctrine restrictively. "That of which we do not see the cause," writes the sixteenth-century Aristotelian doctor and theologian Thomas Erastus, "is not immediately to be considered occult."[9] His contemporary Olmo offers a tripartite distinction between the manifest, the "obscure," and the "absolutely occult"; only that which is "incomprehensible to us," he writes, merits the last designation.[10]

---

[6] *"Stultitae genus videtur, in re, quam occultam dicimus, et revera sentimus, rationem quaerere, occultum enim quid dicere, et rationem quaerere, contradictionem implicant, et egregiam ignorantiam ostendunt."* Giovanni Olmo, *De Occultis in Re Medica Proprietatibus* (Brescia, 1597), p. 47.

[7] Ross, *The new planet no planet, or, The earth no wandring star, except in the wandring heads of Galileans* (London, 1646), p. 76.

[8] Sennert, *De occultis*, p. 55.

[9] *"Non statim occultum est, cuius causam nos non videmus."* Thomas Erastus, *De Occultis Pharmacorum Potestatibus* (Basil, 1574), p. 61.

[10] *"... causam quaedam manifestam obtinet, obscuram alia, alia occultam prorsus, et nobis ipsis incompraehensibilem."* Olmo, *De Occultis in Re Medica*, p. 4.

Sennert, finally, insists that occult qualities are not just the occlusion of a manifest default: "an unknown proportion of manifest qualities," he writes, "does not make occult ones."[11] To some extent, these distinctions may seem consistent with the conceptual triviality, and terminological redundancy, that have traditionally (at least since the seventeenth century) been alleged against scholasticism. Nonetheless, both defenders and *deniers* of the occult-qualities doctrine in the period recognize modes in which phenomena can be causally obscure—hidden, recondite, abstruse—without being, in the strict sense, occult.

The most important of these non-occult modes involves the theory of sympathy and antipathy: the universal doctrine, associated with a Galenic conception of "total substance," that opposites repel and likes attract. For the sixteenth-century Galenist and contagious-disease theorist Hieronymus Fracastorius, the theory of sympathy/antipathy eliminates occult qualities—which Fracastorius finds annoyingly uninformative—from natural philosophy. An iron compass needle, for example, does not point north because of any unique and inexplicable force transferred to it by a magnet. Rather, there are "mountains of iron, and of magnet, under the poles," which distantly attract their likenesses all over the world. It is simply the delicate suspension of the compass needle that allows it to respond to the weak polar force.[12] Of course, if magnetism is just sympathy, the question arises "why a magnet does not attract a magnet more strongly than it does iron; and why iron does not move more strongly toward iron, than toward a magnet."[13] The answer, Fracastorius says, is that the apparent *dis*similarity between the two is deceptive: magnets actually have "another principle, latent within themselves, which is similar either to the iron itself, or to a principle within it."[14] The posit of sympathy and antipathy, in short, explains and even explodes so-called occult qualities.

We find something similar in Jean Fernel—like Fracastorius, a progressive sixteenth-century Galenist. In his *De Abditis Rerum Causis*, Fernel explains contagion by recourse to occult qualities; but he then goes on to explain *away* occult qualities, by recourse to sympathies and antipathies. These are "implanted in everything," he writes, and are "the causes of all the occult results that are

---

[11]   *"Et proportio manifestarum qualitatum ignorata ex manifestis non facit occultas."* Sennert, *De occultis*, p. 57.

[12]   *"Nos igitur diligentius rem considerantes dicimus causam, quod perpendiculum illud ad polum vertatur, esse montes ferri, et magnetis, qui sub polo sunt, vt negotiatores affirmant, quorum species per incredibilem distantiam vsque ad maria nostra propagata ad perpendiculum usque, vbi est magnes, consuetam attractionem facit."* Hieronymus Fracastorius, *In libros de Sympathia et Antipathia rerum*, in *Opera* (Lyon, 1591), p. 28.

[13]   *"Si per similitudinem (vt supra diximus) fit haec attractio, cur magnes non potius magnetem trahit, quam ferrum, et ferrum cur non potius ad ferrum mouetur, quam ad magnetem?"* *Ibid.*, p. 24.

[14]   *"Magnes, et alia multa non trahunt fortasse per id, quod acto sunt, sed per latens aliud in ipsis principium, quod simile ferro est aut ipsi, aut principio in eo."* *Ibid.*, p. 25.

credible on no obvious grounds."[15] But clearly, to nominate a cause for something called occult—a cause, indeed, more intelligible than the "occult result" itself—is to commit exactly the fallacy attacked by scholastic defenders of occult qualities, such as Olmo and Sennert. Indeed, Fernel "tries to throw doubt on the validity" of the occult/manifest distinction, somewhat in the later manner of Descartes.[16] But establishing and securing the validity of the distinction is a non-negotiable priority, as we have seen, for scholastic defenders of the doctrine founded upon it.

To be sure, both Fernel and Fracastorius can validly be called "scholastic." My claim here is not that early-modern Aristotelianism precludes commitment to the theory of sympathy and antipathy. My claim, rather, is that sympathy/antipathy and occult-qualities, within the complex weave of eclectic early-modern scholasticism, are disjunctive, not conjunctive, theories. Belief in one of them is, to some extent, counter-indicative for belief in the other. Some Aristotelians, such as the Cambridge textbook author John Case (who follows Fracastorius on magnetism, and on other abstruse phenomena) apparently hold sympathy/antipathy to the exclusion of the occult-qualities doctrine (which Case barely mentions).[17] Others, such as the late seventeenth-century Wittenberg philosopher Caspar Schön, assert that the two theories overlap, but are not coterminous. "In wine, which is poisonous to many people," Schön writes, "nothing occult is found; neither in the fire that frightens the lion, nor in the mouse that puts the elephant to flight." In short, "not every quality of antipathy and sympathy is among the occult qualities."[18] Erastus takes a similar tack, using the example of cat-phobia. If this were due to an occult quality presented by the species of cats, then cat-phobia would be general to people; just as attraction by magnets is general to iron. But that is not the case. Therefore, the antipathy is not an occult quality.[19] Ross, similarly, notes that people can produce esoteric effects via a "phantasie and prejudicate opinion"; but this is precisely not the same as the operation upon them of "an occult quality."[20]

---

[15]   Jean Fernel, *On the Hidden Causes of Things: Forms, Souls and Occult Diseases in Renaissance Medicine*, trans. John Forrester, eds Forrester and John Henry (Leiden, 2004), p. 49.

[16]   Fernel, *On the Hidden*, p. 25. For Descartes's move, see Hutchison, "What Happened," and Henry, "Occult Qualities."

[17]   John Case, *Lapis Philosophicus seu commentarius in 8 libris* (Oxford, 1599), pp. 716–21. See also Charles B. Schmitt, *John Case and Aristotelianism in Renaissance England* (Montreal, 1983).

[18]   "... *in vino quod multis venenum est ... nihil occulti reperitur, sicut nec in igne qui Leonem terret, in mure qui Elephantem fugat ... Non enim omnes Qualitates Antipatheticae et Sympatheticae qualitatum occultarum sunt.*" Caspar Schön, *Disputatio Physica qua Qualitates Occultas* (Wittenberg, 1665), sigs A4.

[19]   Erastus, *De Occultis Pharmacorum*, pp. 61–3.

[20]   Ross, *Arcana microcosmi, or, The hid secrets of man's body discovered* (London, 1652), p. 83.

Sennert, finally, turns the tables on the Fracastorian view. In the second book of his *Hypomnemata Physica* (1636), Sennert defines what he takes to be the six main sub-categories of occult qualities. The first, "which always coincide with certain species of living things," covers the remora and other fabulous creatures;[21] the third—"of things, which are not alive, which nonetheless have their specific forms, other than from the elements"—covers magnetism.[22] The second, meanwhile, covers sympathies and antipathies. "These properties," Sennert writes, "are not common to the whole species, but are proper to certain individuals … they must be referred to natural strengths and weaknesses arising from a strange tendency of the body, or of a part of it."[23] Sennert's point is not to elide the *prima facie* distinction between sympathy/antipathy and occult qualities— much less make the latter a sub-set of the former (in the manner of Fernel and Fracastorius). Rather, Sennert's point is (1) to *conserve* the *prima facie* distinction between the two kinds of phenomena; and (2) *to make sympathy/antipathy a sub-set of occult qualities*. Like Ross, Sennert makes occult qualities the "causes of *sympathies* and *antipathies*"—whereas the Galenists argued exactly the other way around.[24] Criticizing Fracastorius for relying on a vague notion of magnetic *species* (iterations and projections of a phenomenal essence), Sennert notes that "it is not easy to explain what these immaterial *species* should be." But for exactly that reason, he goes on (*pace* Fracastorius), it is clear that "in the first place they are occult qualities."[25] The whole advantage of occult qualities over sympathies, in this regard, is that they do not require—but reject—explanation. They organize natural philosophy around a fundamental, and ineliminable, strangeness of the natural world.

And what of the supernatural world? Not much. Here again we find a disjunction between sympathy/antipathy and the doctrine of occult qualities. Fernel fulsomely praises, and traces many significant effects, to astrological influence.[26] Case speaks of "the most certain predictions and oracles of Astrology."[27] Levinus Lemnius, a late seventeenth-century adherent of the sympathetic theory, compares magnetic

---

[21] "*… quae viventium quibusdem speciebus semper competunt.*" Sennert, *De Occultis*, pp. 66–7.

[22] "*… rerum, quae non vivunt, quae tamen suas formas specificas, alias ab elementis, habent.*" Sennert, *De Occultis*, p. 69.

[23] "*Hae proprietates cum toti speciei non sint communes, sed quibusdam individuis propriae … ad potentias et impotentias naturales, quae fluunt a peculiari corporis, vel eius partium dispositione referenda sunt.*" Sennert, *De Occultis*, p. 67.

[24] Ross, *The philosophicall touch-stone, or, Observations upon Sir Kenelm Digbie's Discourses of the nature of bodies and of the reasonable soul* (London, 1645), p. 56.

[25] "*Quid autem sint istae species spiritales, non facile explicatu est.*" Sennert, *De Occultis*, p. 77.

[26] Fernel, *On the Hidden*, pp. 33–6.

[27] "*certissima prognostica et oracula Astrologiae.*" Case, *Lapis*, sig. gg4.

sympathy to "the influence of the Starrs ... upon sublunary bodies."[28] Erastus, by contrast, explains the doctrine of occult qualities in part by stating bluntly that "the stars neither produce, nor maintain, the occult properties of things."[29] Olmo, similarly, states against Ficino that the occult qualities of things are not a sidereal function, arguing in evidence that "the magnet, due to its own nature, draws iron at all times and in all places; and its effect is not changed by the changing heavens."[30] Sennert, in his lengthy account of what he reckons to be the six kinds of occult qualities, simply does not include any astral or otherwise cosmic variety among them.[31] Ross, admittedly, is more eclectic: he includes bad air, "infected with the impression of malignant and occult qualities from the influence of the Stars," among the causes of disease.[32] This, however, is but a passing moment in the Scottish Aristotelian's very capacious *Arcana microcosmi*. The latter, in turn, is but one of several books in which Ross has a very great deal to say about occult qualities—without making them, in any significant way, a function of cosmic influence.

Of course, there is one supernatural power that is unquestionably relevant, and quite problematic, for the doctrine of occult qualities: the power of God. Magnetism (*inter alia*) may be occult to us; but it is surely not so to Him. Accordingly, it is conventional in the period to place occult qualities among the "secrets of God," into which it is either foolish or blasphemous to look. The resulting implication, however, is that God is directly and constantly responsible for occult qualities—at least as far as we can tell. All we can say about them, perhaps, is "that God thus willed."[33] But surely this is to say, in natural-philosophical terms, nothing; while suggesting or recalling that there are valid reasons to say the same, and nothing more, of the whole created world. This is Occasionalism: the dark side of Christian (or, more broadly, monotheistic) Aristotelianism, where divine omnipotence

---

[28] Levinus Lemnius, *The Secret Miracles of Nature in Four Books* (London, 1658), bk 4, p. 261.

[29] "*Astra non vel dant, vel conseruant proprietates rerum occultas.*" Erastus, *De Occultis Pharmacorum*, p. 35.

[30] "*Magnes omni ferè in loco, et tempore in omni, ex sui natura trahit ferrum, nec variantibus syderibus variatur effectus.*" Olmo, *De Occultis in Re Medica*, p. 21.

[31] Sennert, *De Occultis*, pp. 66–75.

[32] Ross, *Arcana microcosmi*, p. 83.

[33] "*Satis erit dixisse ... Deum ita voluisse.*" Erastus, *De Occultis Pharmacorum*, p. 37.

cancels out the finite gains of human knowledge.[34] "The study of philosophy is useless," Erastus notes, under such conditions.[35]

Luckily, Erastus, Sennert, Ross et al. are able to deny that occult qualities reduce to divine volition. Instead, they backstop on substantial form. This is the key theoreme of Aristotelian ontology: the irreducible essence or nature of a thing, what makes it what it is.[36] Indeed, the doctrine of occult qualities, for at least some of its neo-Aristotelian defenders in the period, seems to be valuable precisely as a marker of substantial form. At Creation, Sennert writes, God "created all things, and gave to them their forms, making those not idle and ineffective, and deprived of strength; but instructed as to their own qualities and properties and powers."[37] Form determines quality, manifest or occult, as an ordinary and non-Occasionalist part of the way things are. To be sure, there is uncertainty among scholastic theorists as to the exact nature of the form–quality relationship. Erastus is unwilling to see it as simple or straightforward; occult qualities, he writes, are "not in the form, as we are in our house." Nonetheless, qualities or properties "proceed from" their forms.[38] The relationship is key. Ross concludes the *Arcana microcosmi* with a long defense of it.[39] Olmo gives what is perhaps the most acceptable view: occult qualities "neither be form, nor nature; but by nature, and after form."[40] Insofar as occult qualities proceed from the substantial form of a given phenomenon, they are absolutely, even quintessentially, natural, and non-superstitious. They constitute a function of—rather than an exception to—the observable natural order.

Indeed, the scholastic doctrine of occult qualities—and this is a summary point—is first and foremost an *empirical* doctrine. It holds that the world simply contains, as a matter of observational and temporal fact, certain natural phenomena that violate the canons of perception and cognition. Scholastic exponents of the

---

[34]   See Michael E. Marmura, "Al-Ghazali," in Peter Adamson and R.C. Taylor (eds), *The Cambridge Companion to Arabic Philosophy* (Cambridge, 2005), pp. 137–54; Stephen P. Menn, "Metaphysics: God and Being," in A.S. McGrade (ed.), *The Cambridge Companion to Medieval Philosophy* (Cambridge, 2003), pp. 147–70; and Alfred CamFreddoso, "Medieval Aristotelianism and the Case against Secondary Causation in Nature," in Thomas V. Morris (ed.), *Divine and Human Action: Essays in the Metaphysics of Theism* (Ithaca, 1988), pp. 74–118.

[35]   "... *inutile est philosophiae studium.*" Erastus, *De Occultis Pharmacorum*, p. 37.

[36]   See Newman, *Atoms*; David Charles, *Aristotle on Meaning and Essence* (Oxford, 2002); and D. Bostock, *Aristotle: Metaphysics Books Z and H* (Oxford, 1994).

[37]   "... *res omnes creavit, iisque suas formas dedit, easque non ociosas et inefficaces, ac viribus destitutas, sed suis qualitatibus et proprietatibus ac viribus instructas.*" Sennert, *De Occultis*, p. 65.

[38]   "*Proprietates non sunt in forma, ut nos in domo*"; "*a formis proficiscuntur.*" Erastus, *De Occultis Pharmacorum*, p. 20; p. 8.

[39]   Ross, *Arcana*, pp. 255–66.

[40]   "... *neque sit forma, neque sit natura, sed a natura, et post formam.*" Olmo, *De Occultis in Re Medica*, p. 7.

doctrine have seen magnets (to all appearances mere stones) move iron; have undergone notable purgation after eating manifestly innocuous plants; and have heard creditable reports of such incredible creatures as the remora. Moreover—and this point, too, is summary—the qualities that correlate with each of these natural phenomena are caused by those phenomena themselves. It is not just coincidence, for example, that ships suddenly stop dead in the water when proximate to a remora. Neither is this phenomenon to be explained by any other cause—reefs, for example, of which the remora is a mere sign (Fracastorius's view, cited by Case[41]). Rather, the ship-wrecking quality of the remora is totally and directly caused by the remora itself. An occult quality is an *objective* function of the creature from which it arises.

Finally: as empirical and objective violations of perception and cognition, occult qualities cannot be understood. This is where we started. What we have learned, however, is that the unintelligibility of occult qualities is itself an intelligible point within a whole system of intelligibility—the system of scholastic knowledge (*scientia*). A certain dropout from that system is represented by Shakespeare as telling his best friend, still a graduate student, that "there are more things in heaven and earth ... / Than are dreamt of in your philosophy."[42] Horatio, fresh from Wittenberg, might well answer: "Sure. But that means they're irrelevant to (our) philosophy. Also, we're not dreaming." Occult qualities are not a vitiation, but a demarcation, of scholastic knowledge about the natural world. The observational evidence is that one does not know the nature of magnetism (for example); the epistemological fact is that one can never hope to. The resulting doctrine is not just a matter of unanswered questions, but a matter of unanswerable questions. Occult qualities constitute the *undiscoverable country* of early-modern scholasticism.

The Esoteric Inversion

Early-modern esotericism, by contrast, does its best to enter that country. Neoplatonism and Paracelsianism—and alchemy under the sign of either or both—challenge the scholastic doctrine of occult qualities. The esoteric move, however, is to violate the scholastic conception (as it were) by embracing it. Through a combination of misprision and misrepresentation, esotericists re-theorize, to the point of inverting, the scholastic occult.

The process starts from the top. Esoteric theorists of occult qualities (or properties, or virtues) do not limit them, in the manner of the doctrine's scholastic defenders, to the created action of substantial forms. Rather, esotericists refer occult qualities directly to the transcendent and eternally-active will of the Creator. "The occult properties of things," writes the Neoplatonist Cornelius Agrippa, are

---

[41]   Case, *Lapis*, p. 721.

[42]   William Shakespeare, *Hamlet, Prince of Denmark*, ed. Constance Jordan (New York, 2005), 1.5.162–3.

continually "infused from above"; "all vertues," indeed, "are infused by God."[43] His compeer Giambattista della Porta agrees, finding that occult phenomena, rather than being terminal functions of substantial form, "receive their force and power from Heaven."[44] The Paracelsian Richard Bostocke draws the ontological consequences: *arcana naturae* make a mockery of the scholastic idea that God "medleth not under the moon." The fact is that God directs nature all the way down, precisely "in secrete: that is, in power."[45] Of course, it remains the case that God has to work through nature in order to produce His effects therein. This He does, in the Neoplatonic picture, via celestial intermediaries. For Agrippa, "every occult property is conveyed into Hearbs, Stones, Metals, and Animals, through the Sun, Moon, Planets, and through Stars higher than Planets."[46] For della Porta, the natural philosopher needs to consider "the Heavens, the Stars, the Elements, how they are moved, and how they are changed," in order to "find out the hidden secrecies of living creatures, of plants, of metals, and of their generation and corruption."[47] The philosopher, watching the zodiac, should lay matter before the stars "at such a time as such an influence raigneth"; just as a man might "lay iron before the Load-stone to be drawn to it."[48]

Instrumental activity of this kind is natural magic. In the formulation of the Paracelsian alchemist (and opponent of Erastus) Gerhard Dorn, magic "is said to be the fundamental knowledge or understanding of natural things, in the elements of their existence, such as the strengths, virtues, and occult potencies placed and ordered in them by God their creator."[49] In other words, the magus is to *know*, to understand, the nature of occult phenomena. But how is the magus to arrive at this extraordinary knowledge, which clearly violates and exceeds the induction from qualities canonized by more orthodox scholastics? In the only way possible. he gets it from God. Magi, writes Dorn, are a product of "the supernatural heavens" (and not just "the natural stars").[50] According to one of the epistles of the Czech alchemist Michael Sendivogius (in a seventeenth-century

[43]   Agrippa, *Three Books of Occult Philosophy*, trans. J.F. (London, 1651), pp. 35, 28.

[44]   Giambattista della Porta, *Natural magick by John Baptista Porta, a Neapolitane* (London, 1669), p. 10.

[45]   Bostocke, *The Difference between the Auncient Physicke ... and the Latter Physicke* (London, 1585), sigs *****2ᵛ, A3ᵛ.

[46]   Agrippa, *Three Books*, p. 34.

[47]   Della Porta, *Natural magick*, p. 2.

[48]   *Ibid.*, pp. 3, 10.

[49]   "*Magia dicitur scientia sive cognitio rerum naturalium in firmamento, et in elementis existentium, iuxta vires, virtutes, et potentias occultas in ipsis à Domino suo creatore, positas et ordinatas.*" Dorn, *Commentaria in Archidoxorum Libros X, D. Doctoris Theophrasti Paracelsi* (Frankfurt, 1584), p. 450.

[50]   "*Non tamen ... ex naturali stella, sed ex coelo supernaturali prodeunt Magi.*" Dorn, *Commentaria*, p. 453.

manuscript translation), magi result from a process of oral tradition leading all the way back to "Caballistical Principles" "infused by God, into our first parents."[51] The Paracelsian notion of this tradition takes exegetic form: the magus, precisely because he *is* a magus, is able to derive a fundamental physics by allegorically and repeatedly interpreting the creation-narrative of Genesis. The same interpretative authority, argues Dorn, applies to the destruction-narrative of Revelation; from alpha to omega, only the magus, "born or chosen," can legitimately expound God's works. "To the manifestation of secrets, therefore, magic is necessary." The same heavenly powers that produce occult phenomena produce the human minds capable of understanding them.[52]

Thus the magus, via supernatural agency, produces intentional results—real, indeed, precisely because they *are* intentional. His empirical insights are an *a priori* function of his mental and spiritual conformity to God. For that matter, the natural secrets that esoteric practitioners seek to manipulate are themselves intentional. "To the rude," writes a seventeenth-century translator of Dorn, "stonnes and metalles and also minerales the meane betwixte those ... seme not to lyue." Philosophers see differently.[53] "The parts and members of this huge creature the World," writes della Porta,

> do in good neighbour-hood as it were, lend and borrow each others Nature; for by reason that they are linked in one common bond, therefore they have love in common; and by force of this common love, there is amongst them a common attraction, or tilling of one of them to the other. And this indeed is Magick ... Hence it is that the Load-stone draws iron to it, Amber draws chaff or light straws, Brimstone draws fire, the Sun draws after it many flowers and leaves, and the Moon draws after it the waters.[54]

The obverse of the same phenomenon is repulsion of one thing by another. Both Agrippa and della Porta, in their vast and untidy calendars of these effects, cite cat-phobia—the very phenomenon distinguished from occult quality by Erastus. The scholastic, it will be recalled, points out that cat-phobia cannot be an objective quality of the cat, since most people receive no such effect from it. Agrippa, as though parodying that logic, writes that the phobia cannot be a subjective projection of the phobic: "which fear it is manifest is not in them as they are men."[55] What causes things like cat-phobia, for esotericists, is an intentional dynamic between *certain* cats and *certain* people, under certain celestial influences, as perceptible

---

[51] British Library MS Sloane 1800, f. 12ᵛ.

[52] "*Ad manifestationem igitur occultorum, est Magica necessaria ... Vt librum Apolypticae revelationis, nemo nisi Magus natus vel adoptatus, exponere poterit unquam.*" Dorn, *Commentaria*, p. 453.

[53] British Library MS Sloane 1627, f. 3.

[54] Della Porta, *Natural magick*, p. 10.

[55] Della Porta, *Natural magick*, p. 19; Agrippa, *Three Books*, p. 43.

and intelligible by certain favored minds. Such phenomena, moreover, are totally exemplary of the way nature works, from the magus's point of view.

This is the full-fledged esoteric version of the theory of sympathy and antipathy. It is flexible and confident enough to swallow occult qualities whole. Della Porta, echoing Fracastorius, explains that the lodestone (for example) attracts iron because it is stone containing iron.[56] The difference between the sixteenth-century Galenist and the seventeenth-century mage, however, is that the former conceives himself to be debunking occult doctrine by the theory of sympathy/antipathy, while the latter conceives himself to be illuminating it, by the same theory. Sympathy/antipathy comes to be the content, rather than the displacement, of the erstwhile scholastic doctrine. The Jesuit polymath Athanasius Kircher, late in the seventeenth century, attacks the doctrine of occult qualities as "that shameful asylum of ignorance."[57] Like della Porta, and like Fracastorius, Kircher considers magnetism, in particular, to be intelligible as a manifestation of sympathy/antipathy. Kircher, however, goes on to argue that sympathy/antipathy is nothing other than the general manifestation of magnetism. The Jesuit polymath does not, like Fernel, explain *away* the occult quality; he claims to *explain* the occult quality, precisely by making it a basis for further, and yet intelligible, explanations.[58] This is perhaps the ultimate deconstruction of the scholastic view.

Indeed, the fundamental distinction between what I am calling scholastic and esoteric constructions of the occult is that the former entails inexplicability, while the latter does not. Occult can become manifest, according to alchemical, Neoplatonic, and Paracelsian writers. "In the world of the senses," writes Kircher, "there is nothing strictly occult, of which at least a probable cause cannot be assigned." The countervailing view—the scholastic doctrine of occult qualities—moves the Jesuit to a displeasure "that can scarcely be expressed."[59] According to Dorn, "God does not want any secret [*arcanum*] … to be so hidden [*occultum*], but that at any time it might be revealed and made manifest [*manifestum*] by man through magic."[60] Dorn considers the manifestation of the occult to be the most natural process in the world; as a matter of fact, it is *the* natural process *of* the world. "As the rich and secret [*occultum*] seed is pressed by nature into the earth," he writes, "so

---

[56] Della Porta, *Natural magick*, p. 192.

[57] "… *propudiosum illud ignorantiae asylum.*" Athanasius Kircher, *Magneticum Naturae* (Rome, 1667), p. 20.

[58] See Paula Findlen (ed. and intro.), *Athanasius Kircher: The Last Man who Knew Everything* (New York, 2004).

[59] "*In mundo sensibili nil adeò occultum esse, cuius probabilis saltem causa assignari non possit*"; "*dici vix potest, quàm mihi huiusmodi modici animi Philosophi bilem moveant.*" Kircher, *Magneticum*, p. 20.

[60] "*Deus non vult vllum arcanum … tam occultum esse, quin aliquando detegatur, fiatque manifestum homini per Magicam.*" Dorn, *Commentaria*, p. 452.

its secret [*occultum*] appears openly [*in manifesto*] above the earth."[61] As grain produces corn which in turn produces grain, so life revolves via "the occultation of the manifest, and then the manifestation of the occult."[62] Thus where natural philosophy stops, for Erastus, Sennert, Olmo et al., is precisely where it starts for Kircher, Dorn, Agrippa, et al.: with the failure of the senses, necessitating a turn to pure reason, guided by revelation. The occult as undiscoverable becomes the occult as—merely—undiscovered.

Theoretical Implications

The difference between the two positions boils down to a question of what it is *not to know* something. On the doctrine of occult qualities, not-knowing is essentially *nescience*: an encounter with unknowable facts. One does not know certain facts; and one can't; and no-one can, ever. (The question of whether, or how, one can know this—happily—is irrelevant to discussion of the scholastic idea that one does.) Knowability-in-principle is not epistemically requisite to facts as such. Rather, facts divide, epistemically, into the knowable and the unknowable.[63]

If some facts are unknowable, knowledge is finite—explicitly, and by definition. The human mind knows what it can; but it is always confronted and limited, actually or potentially, by the encounter with what it cannot. In the case of magnetism (for example), what one knows is precisely that the nature or cause of magnetism is unknowable. Knowledge in such a case cannot consist in finding out or bringing to understanding the unknown, *qua* unknowable. It can only consist in augmenting and manifesting the understanding of the known. Nescience implies an intensive, rather than an extensive, research program: not the indefinite expansion of factual reference, toward its positable limits; but the indefinite enrichment of factual reference, within limits that have (as it were) always-already been established.

This is consistent with the over-arching scholastic idea of knowledge (*scientia*): "a perspicuous scheme under which to bring together acknowledged phenomena."[64] It is not so much an empirical account of the way things are—what scholastics call knowledge *quia*—as a teleological account of why things have to be the way they are—knowledge *propter quid*. The gold standard of *scientia*, in (many versions of) early-modern scholasticism, is demonstration: a rigorous syllogism, based on essential foreknowledge of its subject-matter, and arguing

---

[61]   "*Vt semen in terram satum et occultum, pellitur à natura sic, vt supra terram occultum eius appareat in manifesto.*" Dorn, *Commentaria*, p. 524.

[62]   "*Generatio dicitur ex occultatione manifesti, mox occulti manifestatio.*" Dorn, *Clavis Totius Philosophicae Chymisticae* (Lyon, 1566), p. 25.

[63]   See Newman, *Atoms*, pp. 139–50.

[64]   Dennis Des Chesne, *Physiologia: Natural Philosophy in Late Aristotelian and Cartesian Thought* (Ithaca, 1996), p. 57.

from necessary premises to necessary conclusions.[65] As Stephen Gaukroger has argued, the scholastic method of demonstration was simply not a method of empirical discovery.[66] The latter is the hermeneutic mode of coming to know new facts; but nescience-*scientia*, as we have just seen, is not fundamentally interested in coming to know new facts. It is fundamentally interested in improving its grasp of established ones.

Now, the hermeneutic corollary of discovery is secrecy. Unknown facts present themselves via the interpretation of data-sets; prior to interpretation, the facts are obscured or occluded (secret) within the data. Interpretation, if successful, discovers the dative secrets, and thereby transforms them into (non-secret) facts. For *scientia*, however, conditioned by nescience, some unknown facts are unknowable. Hermeneutically, that means that some uninterpreted data-sets are uninterpretable. To revert to the traditional metaphor: some secrets are undiscoverable. To be sure, not *all* secrets; but a secret that cannot be found out, clearly, is more secret than one that can. Secrets *per se* are really secret—terminally and inalterably—as a hermeneutic consequence of scholastic epistemology.

This of course is precisely the point of the doctrine of occult qualities. Yet it is easy to miss. Alexander Ross's *Arcana microcosmi* [*the secrets of the microcosm*], for example, is subtitled *The hid secrets of man's body discovered*. Like many in the period, Ross's heading tantalizes with the prospect of occult manifestation. Ross's actual argument, however, is precisely and consistently that the human body, especially in its interaction with pathology, functions in innumerable ways that *must remain secret to us*. Pica (dirt-eating), for example, suggests a digestive power of the stomach which, if physical in the ordinary way, "would doubtlesse corrode the stomach it self"; therefore, "the safest way is to acknowledge an occult quality."[67] Poisons "do not work by their temper which consist of elementary qualities, but by their substance or form, whose qualities are occult to us."[68] Even natural heat, the all-purpose interpretant of early-modern medicine, is for Ross occult: the brain is cold with a coldness that cannot be felt; the lungs are hot essentially, "cold only by accident."[69] In showing "how many strange wonders and secrets are couched up within the Microcosme of our body," Ross precisely does not show secrets, in the sense of revealing them, thereby terminating them.[70]

---

[65]   See Gaukroger, *The Emergence*, pp. 114–48; Robin Smith, "Logic," in Jonathan Barnes (ed.), *The Cambridge Companion to Aristotle* (New York, 1995), pp. 27–65; and Jonathan Barnes, "Aristotle's Theory of Demonstration," in Jonathan Barnes, Malcolm Schofield, and Richard Sorabji (eds), *Articles on Aristotle, 1: Science* (London, 1975), pp. 65–87.

[66]   Gaukroger, *The Emergence*, pp. 159–65.

[67]   Ross, *Arcana*, pp. 141–2.

[68]   *Ibid.*, p. 82.

[69]   *Ibid.*, pp. 15, 30.

[70]   *Ibid.*, sig. a2.

He shows secrets, rather, in the sense of *showing them to be interminably secret*, beyond revelation.

Compare della Porta. The Neoplatonist, like the Aristotelian, "discovers" natural and medical secrets in a compendious book. Porta's *Natural Magick* and Ross's *Arcana microcosmi* might both be called, in period terms, "books of secrets."[71] Yet the similarity between them (so to speak) is superficial. Della Porta does not propose to "discover" secrets in the curious scholastic sense of showing them to be terminally and essentially secret. On the contrary, he proposes to discover secrets in the Neoplatonic sense of essentially manifesting them—thereby terminating, or at least endangering, their status as secrets. The magical art, he writes at the outset of his compendium, "openeth unto us the properties and qualities of hidden things"; "such Things as hitherto lay hid in the Bosome of wondrous Nature."[72] Della Porta then goes on, for hundreds of pages, to give recipe after recipe, treatment after treatment—even, in his hydrostatic and electro-magnetic chapters, parlor trick after parlor trick. To be sure, della Porta rarely gives a detailed account of the *mechanisms* behind his secrets. But that is because, in his estimation, he doesn't have to. He considers the mechanisms of all strange phenomena to be given in advance, as it were, by the theory of sympathy and antipathy. Ross, with his scholastic conception of the occult, would insist that we can never know, even in a vague or general way, how or why a given occult phenomenon is the way it is. Della Porta, with his esoteric conception, thinks he knows this always-already.

And yet—by that very token—the magus is reluctant to show everything he knows. A pose of pedagogic or communicative reticence is absolutely ubiquitous in esoteric writings of the period. In considering whether to publish *Natural Magick*, della Porta writes, "I was somewhat unwilling ... that it should appear to the publike View of all Men."[73] "I had almost forgotten wat silence I am charged with concerning the secretts of nature shutt up in a Shell," wails the author of one alchemical manuscript, fearing to have said too much. "It was not my intent to have written this at all," incontinently states another.[74] Typically, the esoteric author responds to his own problematic power of revealing secrets by adopting an hermeneutics of encoding, characterized by enigmatic utterance and intentional contradiction. Della Porta, for example, says that he has "veil'd by the Artifice of Words" his most wonderful and dangerous secrets.[75] Paracelsus, according to Bostocke, has written "couertly and darkely," even in "the plaine letter."[76] Agrippa, by his own account, generates "Enigmas"—justifying this method by the examples of Plato, Pythagoras, Porphyry, Orpheus, Virgil, the Eleusinian mysteries, the

---

[71]   See William Eamon, *Science and the Secrets of Nature: Books of Secrets in Medieval and Early Modern Culture* (Princeton, 1994).

[72]   Della Porta, *Natural magick*, p. 2; Preface, sig. c1.

[73]   *Ibid.*

[74]   British Library MS Sloane 2218, f. 9^v; British Library MS Sloane 1744, f. 57^v.

[75]   Della Porta, *Natural magick*, Preface, sig. c2^v.

[76]   Bostocke, *The Difference*, sig. L2^v.

apocryphal book of Esdras, and others.[77] Sendivogius, similarly, states himself to have followed the example of the ancient philosophers, who "have everywhere invented fables; made Emblems; and thrown many stones in the way … to hide the Mysteries."[78]

Now, mysteries require hiding only insofar as they have been revealed to those who hide them. Only the knower of a secret knows that he has to keep it. In this way, esoteric rhetoric projects the early-modern magi as having opened up the secrets of nature. It is as though, as Lemnius puts it, the magus's penetration and retailing of secrets might have "furnished Nature with no store."[79] Only and precisely because nature is no longer secret to him must the magus be secretive when talking about it. Furthermore, there is a very strong suggestion in esoteric texts that the decoding of the magus's utterances can never really come to an end. The reader must be patient, attentive, suspicious; he must read the same passages over and over again; he must confer places, and be alert to potentially-significant contradictions. If understanding still does not follow, that may simply be because the reader is not intentionally favored. In other words, he may be too vulgar to understand what he is reading. By that very token, the reader who thinks he *does* understand may need to go back and re-read, in case he has deceived himself. In the end, the secrecy of the esoteric text is, effectively, interminable. The magus does not just seem to possess the secrets of nature; he seems to possess them in an open-ended set.

In a word: *omniscience* is the epistemological precipitate of magistic hermeneutics. The natural-philosophical magi of early-modern esotericism suggest or imply that they know, effectively, all the natural-philosophical facts (once secrets) that there are to know. This is not to say that they really do, or even really think they do. It is to say, rather, that the mode of knowledge the magi project as normative for natural philosophy is determined and defined transcendentally. Whatever the magus actually knows, or does not know, each instance of his knowing is suggestively continuous with, and projects the extremely vast manifold of, the whole and unified realm of facts.

How, then, can *not*-knowing be theorized, on the epistemology of omniscience? Clearly it can have nothing to do with nescience—the theory of unknowable facts. For there *are* no unknowable facts, on the epistemology of omniscience. The omniscient being, by definition, is the being who knows all facts. Knowability-in-principle, from this perspective, is epistemically requisite to facts as such. The difference, accordingly, on the epistemology of omniscience, between a being who knows some facts, and a being who knows all of them, is quantitative, not qualitative. In other words, you can get there (omniscience) from here (knowledge). The knowing of a fact, on this view, is a cognitive procedure which, if it could be repeated *ad infinitum*, would, in principle, produce omniscience.

---

[77] Agrippa, *Three Books*, pp. 556, 346–9.

[78] British Library MS Sloane 1800, f. 69.

[79] Lemnius, *Secret Miracles*, Preface.

The contrast to scholastic *scientia* is absolute. Here, the knowing of a fact is not, by definition, a cognitive procedure that would produce omniscience— not even in principle, not even if it were allowed to run *ad infinitum*. Here, the difference between the being who knows all facts, and the being who knows only some of them, is qualitative, not quantitative. For knowability-in-principle is not epistemically requisite to facts as such. The omniscient being, by that very token, is precisely the being who stands outside *scientia*. To him, all facts are knowable, and he indeed knows all facts. But to the (for lack of a better word) scient being, not all facts are knowable; and he knows, if possible, only those that are. You can't get there (omniscience) from there (*scientia*).

We can construct a theoretical quadrant of correlations and antinomies by way of filling in its final corner—the idea of not-knowing that corresponds to the esoteric idea of the occult. Nescience correlates with *scientia*: the not-knowing of unknowable facts, with the knowing of knowable facts. *Scientia* is the antinomy of omniscience: the epistemology for which not all facts can be known, against the epistemology for which all facts can be known. Clearly, the missing term will correlate with omniscience (for which facts are either known or not known); and will be the antinomy of nescience (the not-knowing of unknowable facts). And this fourth term can only be *ignorance*: the not-knowing of unknown (but knowable) facts.

scientia : nescience

----

omniscience : ignorance

Ignorance is the mode of not-knowing that correlates with omniscience as a standard or idea of knowledge. This is what it is *not to know* something, on the esoteric retheorization of the scholastic occult.

Conclusion

The uncharitable reader may find this result a poor pay-off for the work that led up to it. Nonetheless, the significance of our finding is as follows.

As I have argued, the strict scholastic doctrine of occult qualities is not to be confused with any esoteric theory of sympathy and antipathy. The difference between the two theories may be small, but the vector that emerges from this difference is huge. Occult qualities and sympathy/antipathy trickle down opposite sides of an epistemological watershed. One leads to the idea that some things can never be known; the other, to the idea that everything, in principle, can be known. The ordinary and (to modern eyes) familiar conception of ignorance, moreover, goes with the latter idea, not the former.

As is well known, Isaac Newton recalibrated late seventeenth-century natural philosophy to allow the validity of results based on an occult phenomenon (gravity).

This replaced the dogmatic insistence of Cartesian mechanism that the occult/ manifest distinction was wholly nugatory. According to the canonical narrative of this intellectual history, Newton's settlement of the issue was something like the origin-moment of modern natural science. And so it may have been. The current point, however, is that Newton proffered his natural-philosophical theory of gravity on a basis of ignorance, and not of nescience. He would "scruple not to propose the principles of Motion" in his *Principia*, "and *leave their Causes to be found out*."[80] The idea that the cause or nature of gravity might be *beyond the possibility of finding-out* is remote from, and antithetical to, Newton's thinking about it.

Thus Newton absolutely did not reintroduce to science any idea of occult qualities in the strict scholastic sense. Rather, he kept faith with the idea of sympathies and antipathies in the more modish, esoteric sense. Arguably, therefore, Newtonian physics, as much as Neoplatonic and/or Paracelsian natural magic, leaves open the possibility of human omniscience. The incoherence or questionability of this possibility is what is illuminated by the countervailing early-modern scholastic doctrine of occult qualities. The invention of discovery, as effected by the esoteric magi of the sixteenth and seventeenth centuries, makes knowledge quasi-divine, or not knowledge at all. This rather strange epistemic conception, normalized since by the hegemony of modern natural science, followed from, and depended upon, the end of scholastic nescience.

---

[80]   Cited in Peter Dear, *The Intelligibility of Nature: How Science Makes Sense of the World* (Chicago and London, 2006), p. 37.

Chapter 5
Spirits, Vitality, and Creation in the Poetics of Tommaso Campanella and John Donne

Anthony Russell

In discussions over the meaning of "genius" in the Renaissance, there is at the very least agreement that the term was associated with heightened powers of perception and creativity. Whether defined in terms of Neoplatonic notions of inspiration, Aristotelian/Galenic theories of the humors (and of melancholy in particular), or rhetorical theories of invention, *ingegno* is acknowledged as having denoted an unusual mental energy and flexibility which facilitated the creative process.[1] It is the psychic mobility with which *ingegno* was consistently linked that I intend to emphasize here, and in particular its frequent identification with those volatile spirits that were centrally important in both Aristotelian/Galenic humoral psychology and in the Neoplatonic and hermetic traditions of natural magic. In this respect I am less interested in the early-modern "debate" over the origins of genius between Aristotelian naturalism and Neoplatonic supernaturalism than I am in the continuities between these two very often intersecting traditions.

Though instances of the close relationship between genius, spirits, and creativity can be culled from a plethora of sources from the fifteenth to the seventeenth centuries, I will examine Marsilio Ficino's *De Vita* and Antonio Persio's *Trattato dell'Ingegno dell'Huomo* in order to lay the groundwork for a discussion of poetics in Campanella and Donne. While the *De Vita* first revived an interest in spiritual magic and psychology in the Renaissance, about 100 years later Persio's *Trattato* provided the most sustained study of the pneumatic nature of *ingegno*. In the latter portion of this chapter, I will examine the perspectives on poetic utterance formulated in Campanella's *Poëtica* and Donne's *Anniversaries* in the context of this tradition. Though the *Poëtica* and the *Anniversaries* (both roughly contemporary with Persio's *Trattato*) are very different kinds of works, they exemplify a wider (though by no means universal) trend in which poetry and the process of creation were associated with the inner spirits. Donne and Campanella, I will suggest here, situate poetry somewhere between the conventional alternatives of Aristotelian imitation and divine inspiration by identifying the creative activity

---

[1] See "The Historiography of Ingegno" in Patricia Emison, *Creating the "Divine" Artist from Dante to Michelangelo* (Boston, 2004), pp. 321–48; and Noel Brann, *The Debate Over the Origins of Genius during the Italian Renaissance* (Boston, 2002).

or psychic mobility of "ingegno" as, to some degree, the subject itself of poetic utterance.

If, on the one hand, poetic "invention" entails the process of dis-covering and thus representing a particular objective truth or reality, it is also—perhaps more interestingly—about the activity of creation itself insofar as such activity is closely tied to the creative energies that permeate the cosmos. As a kind of dynamic "exhalation," that is, the poem "embodies" as much as it "represents." This latter conception is of course still a long way from later notions of the radically subjective autonomy of the creative genius; but it may, perhaps, represent a step in that direction.

Spirits, Vitality, and *Ingegno* in Marsilio Ficino and Antonio Persio

"The priests of the Muses," claims Ficino in his *De Vita*, have not sufficiently cared for "that instrument with which they are able in a way to measure and grasp the whole world. This instrument is the spirit, which is defined by doctors as a vapor of blood—pure, subtle, hot, and clear."[2] Though our physical selves are confined and delimited, our pneumatic selves are profoundly permeable to influences of all kinds, since light, color, odors, and voices are all airy, and it is through the inner spirits that we are able even to interact with "the very spirit of the world and with the rays of the stars through which the world spirit acts."[3] The poet, the singer, and the musician, therefore, have a privileged access to the energies that animate the cosmos:

> But remember that song is a most powerful imitator of all things. It imitates the intentions and passions of the soul as well as words; it represents also people's physical gestures, motions, and actions as well as their characters and imitates all these things and acts them out so forcibly that it immediately provokes both the singer and the audience to imitate and act out the same things. By the same power, when it imitates the celestials, it also wonderfully arouses our spirit upwards to the celestial influence and the celestial influence downwards to our spirit. Now the very matter of song, indeed, is altogether purer and more similar to the heavens than is the matter of medicine. For this too is air, hot or warm, still breathing and somehow living; like an animal, it is composed of certain parts and limbs of its own and not only possesses motion and displays passion but even carries meaning like a mind ... Song, therefore, which is full of spirit and meaning—if it corresponds to this or that constellation not only in the things it signifies, its parts, and the form that results from those parts, but also in the disposition of the imagination—has as much power as does any other

---

[2]   M. Ficino, *Three Books on Life*, eds and trans Carol V. Kaske and John R. Clark (New York, 1989), p. 111.

[3]   *Ibid.*, p. 351.

combination of things [e.g., a medicine] and casts it into the singer and from him into the nearby listener. It has this power as long as it keeps the vigor and spirit of the singer, especially if the singer himself be Phoeban by nature and have in his heart a powerful vital and animal spirit.[4]

The poet's or singer's efficaciousness is tied less to abstractive intellectual faculties than it is to those pneumatic faculties (imagination in particular) that yield an immediate connection to the cosmic spirits mediating between the earthly and the celestial, and which are emanations of God's creative vitality. It is the spirit's power, vigor, and mobility that launches it towards the stars, investing song with the power to "heal" both the singer and the listeners.

It is important to note that the concept of creation implicit in the above is neither a conventional theory of imitation, nor a conventional theory of inspiration. If song "imitates" it does so most importantly not because it represents a particular *"res"* or subject-matter, but because it embodies the powers, emotions, and energies of both that which it represents and of the singer. Indeed the primary "matter" of song, in this case, is not "materia" in the typical sense of the word as used in rhetorical theory, but literally a spiritual "substance"—more or less energized depending on the vigor of the singer. Likewise, though in one sense the singer here is inspired, he or she is not simply a mouthpiece of divine *raptus*, but rather, he "arouses the celestial influence downwards to our spirits." In this respect the song is a bridge which permits an audience to sense or to experience the vitality of the universe through the singer's own. Much has been made, rightly, of the importance granted by Ficino to the melancholy temperament as the basis of creative genius, but it is also important to understand that the solitariness of the melancholic thinker has for Ficino less to do with a proto-Romantic embrace of one's alienated subjectivity than it does with the psycho-physiological fact that the melancholic's spirits are concentrated in the brain due to the constant activity of thinking or imagining.[5] Indeed, according to Ficino "the spirits which derive from a melancholy humor are exceptionally fine, hot, agile, and combustible," and this is the reason that such humor is identified with an acute *ingegno* capable of penetrating into hidden realities.[6]

---

[4]   *Ibid.*, p. 359.

[5]   *Ibid.*, p. 111. For a range of approaches to Ficino see D.P. Walker, *Spiritual and Demonic Magic from Ficino to Campanella* (Notre Dame, 1958); Paola Zambelli, *L'Ambigua Natura della Magia* (Milano, 1991); and F. Tomlinson, *Music in Renaissance Magic* (Chicago, 1993).

[6]   "The soul with an instrument or incitement of this kind [the agile spirits]—which is congruent in a way with the center of the cosmos, and, as I might say, collects the soul into its own center—always seeks the center of all subjects and penetrates to their innermost core." Ficino, *Three Books*, p. 121.

As Carol Kaske notes, the *De Vita* was one of Ficino's most popular works. Written in 1480, it went through 30 editions, the last one published in 1647.[7] This may seem an odd fact for a work that was primarily intended as a medical guide for the "priests of the muses," a manual of sorts through which one might learn to achieve and maintain that healthy vitality of the spirit that was essential for the life of the mind. What probably fueled a recurring fascination with this work is its tendency to suggest almost inexhaustible continuities and intersections. Spirit is at the center of these connections: it is the "subtle knot," to borrow a phrase from John Donne's *Extasie*, that links body to soul, earth to the stars, the planets and the sun to human beings, God to humankind, human beings to each other, and so on. It is air, breath, music, voice, light, and life. For followers of Ficino such as Agrippa and Paracelsus, as well for other Renaissance philosophers such as Telesio and Bacon, spirit was the seat of generation and vitality—a kind of cosmic DNA that made possible the cohesion, and variety, of creation.[8]

The close identifications that Ficino established between *ingenium*, imagination, poetic utterance, and spirits in the *De Vita* do not by any means yield a systematic vision of either genius or of poetry,[9] but they do suggest some interesting possibilities which may further complicate our understanding of Renaissance notions of creativity. Most important, perhaps, is the idea that songs or poetic utterances are not simply representational in the conventional sense of the word. In addition to their function as signs conveying a particular meaning or content, the words of song and poetry are also literal exhalations of *spiritus* through which a tangible connection is created between singer, audience, and the celestial spirits immanent in the universe. Song, in this respect, has an animating function. It carries the vital and creative energies that imbue the cosmos. The degree of its animation, moreover, depends on the vitality of the singer's or poet's own spiritual powers. In other words, the song has "power" insofar as it embodies the creative vigor of the singer's own spirits. Medically, song restores health to the listener through a process that we might call re-animation. Thus song occupies a gray area between a kind of "empiricist" psycho-pneumatology, and magic. In one sense Ficino's song is magical, since words have an efficacious power. In another, however, song functions according to the "scientific" logic of spirits, be they bodily or celestial. Words, in this respect, have power insofar as they are uttered and therefore breathed forth as a charged emanation of their singer's pneumatic vigor. Though certainly for Ficino it is important to know what words to speak in the context of the various

---

[7]   *Ibid.*, p. 3.

[8]   M. Fattori and M. Bianchi (eds), *Spiritus. IV Colloquio Internazionale del Lessico Intellettuale Europeo* (Roma, 1984).

[9]   Elsewhere, and in particular in his commentary on the *Phaedrus*, Ficino does discuss poetic frenzy, but in that context he is less concerned with the operations of the human spirits. See Michael B. Allen (trans. and ed.), *Marsilio Ficino and the Phaedran Charioteer* (Berkeley, 1981).

celestial benefices one is connecting to, nonetheless pneumatic vitality seems to be of equal importance in terms of the song's efficaciousness.

Though followers of Ficino such as Agrippa emphasized the magical dimension of his thought, others who were equally interested in the operations of spirits did not necessarily see a need to tie such discussions to magic. So, for example, D.P. Walker has noted that though Francesco Giorgi's *De Harmonia Mundi* is much indebted to Ficino's theories on spirits and celestial influences, it avoids any mention of magic.[10] Like Ficino, however, Giorgi identifies spirit as the locus of creative vitality in the universe as well as in human beings.[11] Perhaps the clearest example connecting human creativity in particular to *spiritus* is Antonio Persio's *Trattato dell'Ingegno dell'Huomo*, a treatise written approximately 100 years after Ficino's *De Vita*, and roughly contemporary to the works of Tommaso Campanella and John Donne. Persio begins his *Trattato* by rejecting the priority of both astrological and temperamental (humoral) causes of genius.[12] Each plays a role in the constitution of genius, but the real seat of *ingegno* is spirit. Spirit is the seat of life and vitality in human beings as well as in the world as a whole, and because "spirit is present to life itself, as its end and without mediation, for this reason it is necessary that its function be to conserve and multiply itself."[13] In human beings, "the more our spirit is similar to that of the sun, the more it will make us ingenious and inventive and judicious, such spirit being as it is entirely luminous, hot, subtle, white, agile, and vivacious."[14] Indeed, it is precisely because the spirit's function is to create and preserve life that Persio views the *ingegno* as pneumatic in nature. It is, after all, through their capacities for inquiry and invention that human beings not only assure their own survival, but continue to improve their lives:

> Because our spirit came from the heavens and became a part of the world, it desires in our bodies every greatness and exaltation, all the more so because due to its subtleness it well perceives its own nobility and excellence, and understands the deficiencies of many things which it would need in its body; and from this sense of need is born desire and appetite, and from this the power of *ingegno* and of human nature are stimulated either by nature or by other men to seek out new things, or to perfect those already found.[15]

---

[10]   D.P. Walker, *Spiritual and Demonic Magic*, p. 112.

[11]   For Giorgi, the world's animation depends on the celestial life which is distributed by the sun "through a certain vital spirit." Human beings receive this vital spirit through their own spirits as well. D.P. Walker, *Spiritual and Demonic Magic*, p. 113.

[12]   Antonio Persio, *Trattato dell'Ingegno dell'Huomo*, ed. Luciano Artese (Pisa, 1999), pp. 30–31.

[13]   *Ibid.*, p. 45. All English translations from the *Trattato* are my own.

[14]   *Ibid.*, p. 40.

[15]   *Ibid.*, p. 47.

In this respect, the operations of *ingegno* are really an aspect of the work of generation and conservation that spirit performs in the universe as a whole. Moreover, the spirit of human beings is particularly worthy of admiration, being both more "noble and more celestial" than the spirit of animals.[16]

Persio's emphasis on the link between vitality, spirits, *ingegno*, and conservation, though already implicit in Ficino, is strongly indebted to Telesio's materialist notion of the soul as a corporeal spiritual substance, and to his notion of conservation understood as the universal instinct to preserve and perfect life.[17] Throughout, however, Persio attempts to reconcile Telesio's sober materialism with the Neoplatonic (and in particular Ficinian) emphasis on love by insisting on the continuities between the lover and the scholar. Persio's emphasis in this regard is on the restless passion that fuels inquiry: "but to proceed further, I say that our spirit, due to its hot nature, always desires to seek out new things, always investigates, always uncovers some occult thing by means of his *ingegno*."[18] It is a passion that revels not only in the specific object of knowledge, but in the *process* of coming to know, a process which in and of itself is pleasurable because it is based in the sensation of the fiery vitality and creativity of the spirit. Thus, for example, Persio offers a fascinating parallel between the union of two illicit lovers, whose commingling of ardent spirits often yields an exceptional child, and Titian's "pneumatic" engagement with his model:

> [Titian] ... according to his own testimony as well as that of those who were present when he was working, when he wanted to draw or paint some figure, having before him a living woman or man, his corporeal sight was so moved by the object and his spirit penetrated so deeply in the model he was depicting, that he seemed to be aware of nothing else, and to the bystanders he seemed to have left his body with his spirit. And it was thought that due to this self-abstraction he was able to create little less than another nature in his works, so well could he express the flesh-tones and the features of the model. The same thing happens between a man and a woman who love each other, spending their best spirits in the process of generation, as also happens with those, who in creating some composition which they weave with animation, imbue it with the highest feelings and elocution that they have.[19]

Insofar as the painting is the "product" of a union between Titian's and the model's spirits, it is, in a sense, a living being. In Persio's examples, the emphasis is placed on creation as a process that has more to do with the "expense" and generation of vitality than on the production of a discrete object. The work that is made—be

---

[16]   *Ibid.*, p. 47.

[17]   On Telesio's view of human psychology, see Neil Van Deusen, "The Place of Telesio in the History of Philosophy," *The Philosophical Review*, 44/5 (1935): pp. 417–34.

[18]   Persio, *Trattato*, p. 59.

[19]   *Ibid.*, p. 69.

it a painting or a text or a human being—is exceptional insofar as it is animated by the vitality of its makers, and the pleasure that it provokes has to do with the quickening of our own spirits. In Persio's words, "a noble *ingegno* becomes happy and joyous in the presence of beautiful persons, because our spirit becomes more subtle and is enlivened in seeing the proportions of a beautiful figure."[20]

Ingegno, in Persio's expansive vision, is much more than the quality of purposive intelligence or creativity or inquisitiveness, though it is these things as well. As the highest manifestation of our spiritual selves, it is the location of our most vibrant sense of vitality; it is most proximate to that *"spirito universale"* that binds and enlivens and conserves all of creation; it is also the origin and the medium of our desire to seek out and to unite with that which perfects us, be it the occult mysteries of nature, the beauty of art, a person's virtue, or the angels of heaven.[21] To practice creativity is in itself, besides the utility of particular discoveries or achievements, to align oneself with the vitality of God's universe.

The Poetics of Vitality in Tommaso Campanella and John Donne

Tommaso Campanella's theory of poetry grows out of the pansensism that is central to his writings in natural philosophy. Like Antonio Persio, Campanella was deeply indebted to Telesio's thought, and in *Del Senso Delle Cose e della Magia* he elaborated upon the latter's notion of the world as sensate by privileging, like Ficino and Persio, the pneumatic connectedness of all creation. "The world," he announces immediately under the title of his work, "is a living and sentient statue of God, and all its parts and particles are conscious—some more some less—to the degree necessary for their own conservation and for the conservation of the larger whole in which they are consentient."[22] All creation is alive and sensate because all beings, animate and inanimate, are ensouled by a "hot, subtle, and lively spirit, capable of impression and immediate sensation, like the air."[23] Though Campanella allows that human beings are unique in possessing a mind that is directly infused by God, he is generally more interested in exploring the implications of the *"consenso"* or "conspiracy," by which all creatures are connected.[24] Rather than taking the slow and indirect route of discursive logic and cognitive abstraction, Campanella argues that we can achieve a more tangible and

---

[20]   *Ibid.*, p. 76.

[21]   *Ibid.*, p. 82.

[22]   Tommaso Campanella, *Del Senso delle Cose e della Magia*, ed. Germana Ernst (Bari, 2007), p. 1. All English translations of Tommaso Campanella's works are my own.

[23]   *Ibid.*, p. 12.

[24]   I am borrowing the evocative term "conspiracy" from Crashaw's hymn "To the Name Above Every Name, the Name of Jesus," in which the poet calls on all of creation to contribute to his song: "Bring All the store / Of *Sweets* you have; And murmur that you have no more. / Come, nére to part, / *Nature & Art*! / Come; & come strong, / To the conspiracy

more compelling knowledge of God and of his creation by accessing those vital energies that link us directly to the world as "sensation and life and soul and body, statue of the Most High."[25] Campanella insists, for example, on the etymological relationship between "*sapere*" (knowledge) and "*sapore*" (taste) to convey the sensual immediacy that his mode of inquiry entails.

Often, moreover, Campanella seems to have been less interested in the practical applications of this kind of approach to reality than in its moral benefits. In the epilogue to *Del Senso delle Cose*, Campanella offers a striking evocation of the moral imperative of his cosmology:

> Like worms within an animal are all animals within the World; nor do they imagine that the World might be sensate, just as the worms in our stomach do not think that we feel and have a soul greater than theirs ... Blessed he who reads in this book and learns of the nature of things from it, and not from his own whim, and learns too the divine art and government, and consequently makes himself similar to and at one with God, and sees that everything is good, and that the bad is relative ... And he thus takes pleasure in, admires, reads, and sings the infinite, immortal God.[26]

Campanella exhorts us to perceive, indeed to *feel*, the common "sense" shared by all of creation. By so doing, we recognize ourselves as part of a larger, vital whole made and animated by God, and through this recognition we will realize that the deepest fulfillment of our instinct for conservation—for the preservation of our lives—lies in our joyous participation in the harmony of God's creation, a participation that yields a vision of mutability as encompassed within the embrace of eternity. Campanella's natural philosophy is, therefore, at the same time moral philosophy, because it calls for an awakening of the senses, a reorientation of perspective through which knowledge of ultimate truths turns out to be something that is immediately available, and can be directly tasted, rather than abstractly conceived.

In the *Philosophia Rationalis*, the most ambitious and extensive exposition of his ideas, Campanella grants poetry the highest status among the disciplines,[27] and, in a fairly provocative move, he defines the poem as a "magical instrument,"[28]

---

of our Spatious song (ll. 66–71)." In George W. Williams (ed.), *The Complete Poetry of Richard Crashaw* (New York, 1970), p. 30.

[25]   Campanella, *Del Senso*, p. 235.

[26]   *Ibid.*, pp. 235–6.

[27]   Campanella, *Poëtica*, in *Tutte Le Opere di Tommaso Campanella*, vol. 1, ed. Luigi Firpo (Verona, 1954), p. 1028.

[28]   *Ibid.* p. 1017. On Campanella's poetry see especially Pasquale Tuscano, *Poetica e Poesia di Tommaso Campanella* (Milano, 1969); Franc Ducros, *Tommaso Campanella poète* (Paris, 1969); Dennis Costa, "Poetry and Gnosticism: The *poetica* of Tommaso Campanella," *Viator: Medieval and Renaissance Studies*, 15 (1984): pp. 405–18; and

"the most perfect manifestation of vocal magic."[29] According to Campanella, the feeling of the "perfect conservation of our own life" is the greatest pleasure that human beings can experience, and the magic of poetry is most capable of arousing such sensation.[30] This magic works in three ways. In the first place, poetry's meter replicates the natural rhythms of our inner spirits. As our spirit comes into contact with the pneumatic pulsations of metric utterance, it experiences a sense of purification, expansion, and animation; it literally experiences its own conservation through the renewal of its vitality. In the second place, Campanella locates the power of poetry in its imitative function. He rejects, however, the Aristotelian theory of mimesis, by locating the pleasure felt in a representation not in the experience of recognition of the object depicted, but in our admiration of the skill of the artist as it vividly reminds us of the "*potentia*" (power) and "*sapientia*" (wisdom) of humankind through which we conserve and perpetuate ourselves.[31] Poetic skill, in other words, manifests the creative capacities through which human vitality is preserved.

The third aspect of the magic of poetry is that it speaks directly to the senses: "Poetry is a reversed form of science, because it first speaks to the will, not to reason, as do the other sciences."[32] By speaking directly to the senses, it is more likely to transform its audience: "Because poetry is the most perfect manifestation of vocal magic, it moves souls more powerfully … by presenting objects to the feelings and communicating its own emotions."[33] Finally, in keeping with his emphasis on tangibility, Campanella also advises the poet to avoid the creation of fables and to engage rather with the real world insofar as it is the living statue of God. The noblest forms of poetic expression are the psalmic praise of God and of his creation, and the philosophic or scientific poem.

Readers who turn to Campanella's poetics intrigued by citations about poetry as a "magical instrument" may find themselves disappointed by how limited in some ways the claims for this magic really are. Setting aside the effects of meter on spirit, Campanella's description of the power of poetry does not seem so different from the universally accepted rhetorical theory of the *affectus*, whereby eloquent speech moves the emotions of human beings. Indeed, though the first version of his poetics (1596) is substantially similar to the second (1612–38), nowhere in the first does he use the term "magic." So why did Campanella insist on including in the second version a term that, in the climate of the Counter-reformation, may have seemed provocative? It may be useful in this context to turn to a little noticed definition of magic that Campanella provides in *Del Senso*:

Marziano Guglielmetti, "Magia e Tecnica nella Poetica di Tommaso Campanella," *Rivista di Estetica*, 9 (1964): pp. 361–400.

[29]   Campanella, *Poëtica*, p. 1017.

[30]   *Ibid.*, p. 317.

[31]   *Ibid.*, p. 921.

[32]   *Ibid.*, p. 911.

[33]   *Ibid.*, p. 1017.

> Everything that scientists do in imitation or aid of nature through unknown arts, is called a magical operation, not only by lowly peasants, but even by civilized men ... While the art is not understood, it is called magic: after, it becomes vulgar science.[34]

Magic, in other words, is simply an art or discipline—operating through the pneumatic permeability of the world—whose procedures are not understood *yet*, and whose effects thus simply *seem* miraculous. From Campanella's pansensistic perspective, of course, this follows logically. If all things are alive and interconnected by spiritual means, then all actions and operations in the world must be understood in terms of this dynamic "consent." As he states in the *Poëtica*, the growth of harvests from a few seeds is greater magic than the multiplication of five loaves and two fish to feed 5,000 people, but since the latter is a rare event, it is more admired.[35] The reason Campanella defines poetry as *magia*, I believe, is that his real concern is to resist what pulls us away from a feeling of imbrication—what leads us to feel separate or distinct from God's living statue. The poet, by possessing a particularly dynamic and subtle spirit through which he penetrates into the vitality of all things, by reviving through meter the very motions of the spirits occulted within us, and by choosing to be grounded in the truths of nature and of our lives, infuses in his reader a sensation, a taste (*sapere-sapore*) of his own vital and productive connection with humankind, with nature, and with the processes through which the world, God's living statue, continues to be animated.[36] From this perspective our abilities, our arts, are always-already magical. Poetry is magical not because it *does* something out of the ordinary, but because it yields in us a *sense* of the ordinary as extraordinary, that is, of greater moral efficacy than any abstract dialectical discourse. Its pleasures are at once aesthetic, ethical, and physiological.

On first sight, John Donne's gloomy account of the world's atrophy in the *Anniversaries* would seem to have little to do with Campanella's affirmation of the world's vitality and of poetry's capacity to realign us with it. Yet these works do not simply reduce to a *contemptus mundi* motif. That Donne conceived of poetry as a particularly effective means of projecting its speaker's forcefulness and charisma has been readily recognized, but in the *Anniversaries* he provided a much more intense and sustained reflection on poetry's "powers" which yielded some unique claims.[37] These claims, I will suggest, are closely linked to his perception of the world in these poems in terms of spirits and vitality.

---

[34]   Campanella, *Del Senso*, p. 241.

[35]   Campanella, *Poëtica*, p. 1037.

[36]   See *ibid.*, p. 911, in which the author compares human procreation to poetic making.

[37]   Among readings of the *Anniversaries* that focus on Donne's reflections on the nature of his own verse, see Anthony F. Bellette, "Art and Imitation in Donne's *Anniversaries*," *SEL*, 15/1 (1975): pp. 83–96; Peter L. Rudnytsky, "'The Sight of God': Donne's Poetics

No doubt the most famous passage from the two elegies is the one from the *Anatomy* that associates the decay of the world with the rise of what Donne calls "new Philosophy":

> And new Philosophy cals all in doubt,
> The Element of fire is quite put out;
> The Sunne is lost, and th'earth, and no man's wit
> Can well direct him where to looke for it.
> And freely men confesse, that this world's spent,
> When in the Planets, and the Firmament
> They seeke so many new; they see that this
> Is crumbled out againe to his Atomis.
> 'Tis all in pieces, all cohaerence gone. (*FA* ll. 205–13)[38]

"New philosophy" calls into question traditional beliefs in the organic cohesiveness of the universe. The vital unity of the world gives way, under the anatomizing gaze of scientists and astronomers, to a sense of the world as inert matter with no special place in a randomly ordered cosmos. In the context of this loss of corporate identity, "euery man alone thinkes he hath got / To be a Phoenix, and that then can bee / None of that kinde, of which he is, but hee" (*FA* ll. 216–18). Willfully-affirmed distinctiveness becomes the only ground for identity. We are reminded of the contrast Campanella draws between those "blessed ones" able correctly to read in the book of nature the *consenso* among all things, and those whose perspective on reality is determined by "their own whims." In this context, Donne imagines the meridians and parallels by which the astronomer measures the heavens as a "net throwne / Upon the heauens, and now they are his owne. / Loth to goe vp the hill, or labour thus / To go to heauen, we make the heauen come to vs" (*FA* ll. 278–82).

It is important to note that nowhere in these poems does Donne question the objective claims of the "new Philosophy." What concerns him more than the empirical accuracy of its descriptions of the cosmos is the isolated and morally impoverished sensibility that it yields. In the primary conceit of this elegy, of course, it is Elizabeth Drury's death and absence that yields such a barren world, but in the specific context of scientific discourse, the forsaken alternative to "new Philosophy" is described in a less frequently noticed passage on natural magic:

---

of Transcendence," *Texas Studies in Literature and Language*, 24/2 (1982): pp. 185–207; and Anthony Russell, "'Thou seest mee strive for life': Magic, Virtue, and the Poetic Imagination in Donne's *Anniversaries*," *Studies in Philology*, 95/4 (1999): pp. 374–410.

[38]   All citations of Donne's poetry are from Gary A. Stringer (ed.), *The Variorum Edition of the Poetry of John Donne*, vol. 6 (Bloomington, 1995) pp. 2–37. Citations by line number in the body of my text.

> What Artist now dares boast that he can bring
> Heauen hither, or constellate any thing,
> So as the influence of those starres may bee
> Imprisond in an Herbe, or Charme, or Tree,
> And doe by touch, all which those starres could do?
> The art is lost, and correspondence too.
> For heauen gives little, and the earth takes lesse,
> And man least knowes their trade, and purposes. (*FA* ll. 391–8)

The "artist" here is the magus-astrologer whom Donne imagines as no longer able, in this "spent" world, to access those hidden virtues and influences that are the wellspring of creative vitality throughout the cosmos.[39] The deadness of the world, in other words, is associated here with a loss of connection to those vital spirits that guaranteed the continued coming into being of the world—spirits that could be in various ways channeled or manipulated by those who perceived and understood the complex network of correspondences between the heavens and the earth.

The nostalgia or regret expressed here does not of course necessarily reflect Donne's belief in the operative claims of natural magic. The image of the artist "boasting" even hints at the skepticism Donne elsewhere expressed about some of these claims (in *Love's Alchemy* for example). Indeed, Donne's lament that "the art is lost, and correspondence too" inconsistently blames both the magical operators (who have forgotten their craft), and the cosmos they operate in (which is lacking in correspondences). If the poem describes a world whose objective reality does not allow "commerce twixt heauen and earth," then no amount of "recollection" on the part of esoteric practitioners can remedy this. Whether the loss described is based on objective facts or a failure of imagination, the poet's chief preoccupation seems to be the *perspective* on reality that we are left with in a post-magical context, a perspective from which the cosmos is no longer viewed as a living, organic, and therefore divinely constituted whole. This is the principal reason for which the world is declared dead in this sequence of poems.

Indeed, the figurations that describe Elizabeth *in vita* in the *Anatomy* insistently allude to the occult powers or virtues that were of central importance in the traditions of natural magic from Ficino to Paraclesus and Agrippa. Elizabeth's virtue is imagined as something much more tangible and vibrant than an abstract ideal. It is a dynamic pneuma or spiritual force that integrated the cosmos and thus preserved the vital link between God and humankind. Donne describes her, using Paracelsian terminology for the vital spirits occulted in nature, as an "intrinsique Balme" and "preservative" (*FA* l. 57). He also refers to her several times as a celestial influence or emanation (*FA* ll. 378, 415) whose vital essence "did inanimate and fill / The

---

[39]   See John Freccero, "Donne's 'Valediction: Forbidding Mourning,'" *ELH*, 30 (1963): pp. 335–76; and Eugene R. Cunnar, "Donne's 'Valediction: Forbidding Mourning' and the Golden Compasses of Alchemical Creation," in L. Frank (ed.), *Literature and the Occult: Essays in Comparative Literature* (Arlington, 1977), pp. 72–110.

world" (*FA* ll. 68–9). She was an agent of alchemical sublimation whose "vertue ... so much refin'd" drove out the "poysonous tincture" of original sin (*FA* ll. 177–82); she was a "Magnetique force" of sympathetic attraction "that should all parts to reunion bow" (*FA* ll. 221–2). Her virtue was "the Cyment which did faithfully compact / And glue all vertues, now resolv'd, and slack'd" (*FA* ll. 49–50). She was the principle of life itself, "from whom / Did all things verdure" (*FA* l. 364). At her death, Donne claims, the "vital spirits" of the world were dissipated (*FA* l. 13).

As John Carey and others have noted, throughout his writings Donne tended to seek an imbrication of the physical and the spiritual.[40] He was consistently fascinated with that "subtle knot"—the pneumatic juncture of the corporeal and the incorporeal—that, as he puts it in the *Extasie*, "makes us man" (l. 64). The *Anniversarie* poems confirm this view. Though their trajectory takes us from the lifeless corpse of the world to Elizabeth's swift flight to heaven, Donne's imaginative focus remains insistently on Elizabeth as the link between earth and heaven, as the vital "balsamum" or "preservative" that endowed the world with life. In Elizabeth's absence, Donne invests his own poems with the integrative role she performed. By poetically embodying Elizabeth, he tells us later (in the *Second Anniversary*), the world may be "embalmed" and "spiced" (*SA* l. 39). "Balm," in Paracelsian philosophy, is another term for the "astral body" or principle of vitality occulted in all things, and indeed Paracelsus argues that even corpses preserve some of this vitality, since in his essentially dynamic vision of matter there is no such thing as absolute lifelessness.[41] By associating his poetic utterance with Paracelsian "balm," Donne endows it with the task to "conserve" life. As in the case of Campanella, poetry embodies a vital and generative energy that is the currency of the "commerce twixt heauen and earth" (*FA* l. 399).

We should remind ourselves, at this point, that Elizabeth was in fact a nobody. She was a 14-year-old girl who, as Ben Jonson pointed out, did not in objective terms merit the praise she received.[42] To remember Elizabeth is therefore really to will a sign into existence that will fill the gap opened up by the natural magicians. Donne takes over the role of the magician in restoring the "commerce twixt heauen and earth," but he does so in a very different context, a context in which traditional magical correspondences are in doubt. What these poems do, therefore, in "creating" Elizabeth, is to preserve or "conserve" our sense of the divine vitality

---

[40]   See Freccero, "Donne's 'Valediction: Forbidding Mourning'"; John Carey, *John Donne: Life, Mind, and Art* (New York, 1981), pp. 131–67; Terry Sherwood, *Fulfilling the Circle* (Toronto, 1984); and Thomas Docherty, *John Donne, Undone* (New York, 1986).

[41]   Owen Hannaway, *The Chemists and the Word* (Baltimore, 1975) pp. 23–31; and Don Cameron Allen, "John Donne's Knowledge of Renaissance Medicine," *The Journal of English and Germanic Philosophy*, 52/3 (1943), pp. 340–41.

[42]   Ben Jonson, "Ben Jonson's Conversations with William Drummond of Hawthornden," in C.H. Hereford and P. Simpson (eds), *Ben Jonson* (Oxford, 1925), p. 133.

inherent in human beings in the face of mortality, contingency, and epistemological uncertainty.

In the last lines of the *Anatomy*, Donne affirms that "verse hath a middle nature" (*FA* l. 473) whose function is to preserve the memory of Elizabeth, while the earth keeps her body, and heaven her soul. It is poetry, for both Campanella and Donne, that most effectively bridges the gap between earth and heaven, and perhaps also between magical and non-magical conceptions of language. Just as Campanella sees poetic utterance as having the unique power to pluck at the chords of life and generation, so Donne affirms the generative power of his verses: "These Hymes may worke on future wits, and so / May great Grand-children of thy praises grow" (*SA* ll. 37–8). For both poets, there is something crucially at stake in "animistic" perceptions of the vitality of the world. If, as Donne acknowledges, the concrete claims of magic may not remain viable in light of the "new Philosophy," nonetheless its conception of the world and its creatures as intimately linked to their divine maker is of crucial moral value, and it is up to the poet to preserve this conception through the creative process itself. The poet, in this context, is neither Aristotelian imitator, nor inspired mouthpiece of God. Rather, the act of poetic making becomes an individually willed participation in the vitality which is God's continued gift to us, however we choose to perceive the universe we inhabit. What the poet dis-covers, through the inventive activity that produces his poem, is the redemptive creative energy that makes such invention possible.

Chapter 6

Perfection of the World and Mathematics in Late Sixteenth-Century Copernican Cosmologies

Pietro Daniel Omodeo

The sixteenth-century astronomical debate began with Nicolaus Copernicus's discovery of the planetary heliocentric (or rather "heliostatic") system.[1] After the first reception of this work by German mathematicians, who were primarily interested in the calculation of planetary positions,[2] the debate focused on general hypotheses and geometrical models, on cosmology and natural philosophy, and also involved the epistemological status of mathematics. This comparative survey of three influential Copernicans—the mathematician Johannes Kepler, the pre-Galilean physicist Giovanni Battista Benedetti, and the philosopher Giordano Bruno—is aimed at reflecting on the role of philosophical and epistemological assumptions for the development of a new mathematical science. Moreover, it will shed new light on the so-called Copernican or astronomical revolution, questioning the historical commonplace that Copernicanism was a standard world view, rather than the eclectic composition of diverging cosmologies. The work of Benedetti, relatively unknown to scholars, will be particularly helpful in illuminating just what it was to learn and know in the context of post-Copernican cosmological debate.

Empirical Astronomy

In his *Mysterium cosmographicum* (Tübingen, 1596)—"a work ... of tiny bulk, of modest effort, of contents in every way remarkable"—Kepler revealed the archetypal reasons for heliocentrism, which he asserted to have been part of the

---

[1]   This "heliostatic" aspect of the Copernican system was highlighted by E. Rybka, "Kepler and Copernicus," in A. Beer and P. Beer (eds), *Kepler: Four Hundred Years. Proceedings of Conferences Held in Honour of Johannes Kepler* (Oxford, 1975), pp. 209–16; and O. Gingerich, "Kepler's Place in Astronomy," in Beer and Beer, *Kepler*, pp. 261–8.

[2]   See R.S. Westman, "The Melanchthon Circle, Rheticus and the Wittenberg Interpretation of the Copernican theory," *Isis*, 66 (1975): pp. 163–93.

secret doctrines of the Pythagoreans.[3] This attempt at a metaphysical foundation of the Copernican system opposed the empirical approach of Kepler's immediate forerunners, who, particularly in the 1570s and 1580s, concentrated on the observation of the heavens and the recording of new data. Copernicus had already taken a certain number of accurate observations for the improvement of celestial parameters. In the *De revolutionibus*, Copernicus also described his instruments (similar to those of Ptolemy): the solar quadrant, the armillary or spherical astrolabe, and a "parallactic instrument" called the *triquetrum*.[4] The empirical ground of *De revolutionibus* was very much appreciated in the early-modern period. Sometimes it was even overemphasized, as in the case of Giovanni Antonio Magini, professor of mathematics at the University of Bologna, who assured that his ephemerides agreed with "Copernicus's observations."[5] Actually, they relied merely on Erasmus Reinhold's Copernican tables (*Tabulae prutenicae*, Tübingen, 1551).[6]

Other members of the generation before Kepler also prosecuted the project to develop astronomy through records of new empirical data. In Kassel, the Landgrave Wilhelm, patron of astronomy, charged his skilful craftsman Jost Bürgi with building observational instruments and planetary models. Moreover, he worked together with some of the major astronomers of the time (Rothmann, Brahe, Flamløse, Wittich, Ursus), who visited him or resided at his court. He exchanged data and opinions with Brahe through an intense correspondence published in 1596.[7] Brahe himself was a keen celestial observer: his observatory of Uraniborg is one of the technical marvels of sixteenth-century science.[8] He published the images and the description of his instruments in a work titled *Mechanica* "for the restoration of astronomy" (Wandesburg, 1598). Later, Brahe's data enabled Kepler to perfect his celestial geometries, to discover the laws of planetary motion, and to compute the *Rudolphine Tables* (Ulm, 1627).

Mästlin, the master of Kepler in Tübingen, shared the conviction that astronomy should rely on the precise recording of data. In *Ephemerides novae* (Tübingen, 1577), he urged astronomers to begin with celestial observations instead of intellectual or abstract speculations.[9] In order to measure the elevation of sun and

---

[3]   J. Kepler, *Mysterium cosmographicum*, in *Gesammelte Werke*, ed. by M. Caspar (Munich, 1938), pp. 9–14, p. 5.

[4]   Copernicus, *De revolutionibus* II,2; II,14 and IV,15.

[5]   G.A. Magini, *Ephemerides Coelestium motuum ... secundum Copernici observationes ... supputatae* (Venice, 1609).

[6]   G.L. Betti, "Il copernicanesimo nello Studio di Bologna," in M. Bucciantini and M. Torrini (eds), *La diffusione del copernicanesimo in Italia. 1543–1610* (Florence, 1997), pp. 67–82.

[7]   T. Brahe, *Epistolarum astronomicarum libri*, in *Opera Omnia* (Amsterdam, 1972), vol. 6.

[8]   J.R. Christianson, *On Tycho's Island: Tycho Brahe and His Assistants, 1570–1601* (Cambridge, 2000).

[9]   M. Mästlin, *Ephemerides novae* (Tübingen, 1580), ff. X3v–X4r.

planets, and the distances of the celestial bodies, he indicated two instruments: a *quadrans magnus* and a *radius*. Like Wilhelm of Hesse and Brahe, Mästlin planned to perfect astronomy, in particular its predictive part, which he called *astronomia practica*. He adhered to Copernicus's hypotheses, though he preferred not to write openly about them.[10]

Nonetheless, in the astronomical disputation *De astronomiae hypothesibus sive de circulis sphaericis et orbibus theoricis* (Heidelberg, 1582), Mästlin dealt with an epistemological issue. He gave a solution to the problem of the real (material) existence of astronomical circles and orbits, placing them between conventionalism and realism.[11] Celestial orbs, he wrote, are deduced *a posteriori*, and not *a priori*, because no one has privileged access to the ethereal region. Therefore, the names of the celestial zones are partly conventional. Yet they correspond to something real, as for instance "orient" refers to the place where the sun rises, and "ecliptic" refers to its path.

The Cosmological Turn

Reflection on the natural and physical consequences of the new astronomy matured in the 1590s. This "conceptual revolution," regarded by Koyré and Kuhn as the "astronomical" or "Copernican" revolution *tout court*,[12] received a decisive impulse from discussion about geo-heliocentric world systems. Such models were invented in order to combine Copernicus's geometry (as Kepler saw it, the "unity" of the celestial phenomena based on a new conception of the relationship earth–sun–other planets),[13] with Aristotle's physics (which implied the centrality and immobility of the earth). Granada has convincingly dated the "cosmological turn" back to 1588,[14] when two printed books presented for the first time the geo-heliocentric model: Ursus's *Fundamentum astronomicum*, and Brahe's *De mundi aethereis recentioribus phaenimenis*.[15] Also in 1588, Bruno printed a Latin treatise

---

[10]   *Ibid.*, ff. X4v–XX1r: "*[Divina bonitas] Copernicum, alterum Ptolemaeum, hypothesium observationibus congruentissimarum prolatione decoravit.*"

[11]   M. Mästlin, *De astronomiae hypothesibus* (Heidelberg, 1582), f. A2r.

[12]   A. Koyré, *The Astronomical Revolution: Copernicus-Kepler-Borelli*, trans. R.E.W. Maddison (Paris, London and New York, 1973); T. Kuhn, *The Copernican Revolution* (Cambridge, MA, 1957).

[13]   J. Kepler, *Astronomia nova*, in *Gesammelte Werke* (Munich, 1937), pp. 8–35, p. 9.

[14]   M.A. Granada, *El debate cosmológico en 1588. Bruno, Brahe, Rothmann, Ursus, Röslin* (Naples, 1996).

[15]   To be precise, the first elements of geo-heliocentrism are to be found in the manuscripts of Erasmus Reinhold and Paul Wittich, as well as in K. Peucer, *Hypotheses Astronomicae* (Wittenberg, 1571). See O. Gingerich and R.S. Westman, "The Wittich Connection: Conflict and Priority in Late Sixteenth-Century Cosmology," *Transactions of the American Philosophical Society*, 78 (1988).

directed at German scholars, *Acrotismus comoeracensis*, in which he described and argued for an infinite universe containing an infinite number of solar systems. He and other supporters of Copernicus needed to elaborate answers, both physical and philosophical, to the geo-heliocentric challenge. This confrontation forced them to deepen the epistemological problem of the status of mathematics in natural science. A reflection on the relationships of hypotheses to nature, and mathematics to physics, was necessary in order to demonstrate the plausibility of the Copernican model: that is, to establish whether it was just a skilful "invention," or rather a "discovery."

Kepler's *a priori*

Kepler, however, did not clearly distinguish between invention and discovery. As is well known, Kepler believed that there were binding geometrical reasons, founded in God's mind, for heliocentrism. He found a surprising correspondence between the five regular solids (the so-called "Platonic solids"), and the six planets. In fact, he succeeded in inscribing and circumscribing the celestial spheres in these geometrical figures, in full observance of the astronomical distances. He regarded this remarkable coincidence as the intelligible proof of cosmic harmony and Divine Providence.[16]

Mästlin was first informed of this *ratio a priori* in a letter dated October 3, 1595. He appreciated Kepler's *inventio* so much that he took upon himself the responsibility of publishing the *Mysterium cosmographicum*, together with a new edition of Rheticus's *Narratio prima* and his own calculations of planetary distances from the center of the world (a kind of verification of Kepler's model). Though Mästlin had earlier considered aprioristic astronomy impossible, in the preface to the *Mysterium* he approved his pupil's attempt to deduce the planetary theory from archetypal principles ("*a fronte*"), instead of from the effects ("*a terga*"), in contradistinction to his predecessors.[17] Mästlin thought that Kepler's metaphysical discovery/invention would lead to universal acceptance of Copernicus's teaching.

---

[16]   See R. Martens, *Kepler's Philosophy and the New Astronomy* (Princeton and Oxford, 2000); and P. Barker and B.R. Goldstein, "Theological Foundations of Kepler's Astronomy," *Osiris*, 16 (2001): pp. 88–113.

[17]   M. Mästlin, *Candido Lectori*, in Kepler, *Mysterium*, pp. 82–5, p. 82: "*Astronomiam tamen hactenus omnes non nisi a tergo adorti sunt, et tam motus, quam magnitudines et distantias ex Solis observationibus indagare docuerunt. An autem a priori, sive a fronte ullus ista dimetiendi pateat aditus, vel anne ulla alia, praeter observationes, geometrica Norma, inventos motuum et quantitatum numeros examinandi, haberi possit, nulli ne peritissimo quidem Artifici hactenus, vel per insomnium, in mentem venit.*"

On exactly this basis, his adherence to the heliocentric hypotheses became fervent.[18]

To be sure, Kepler also investigated the physical causes of planetary motions. In the twentieth chapter of the *Mysterium*, he considered the sun as the "universal motive soul," cause of all planetary motions through distant action. He regarded it as the luminous image of the first person of the Holy Trinity, and called it also "world hearth," "king," "emperor," and "visible God." Kepler considered remoteness from the centre as the reason for the different periods of the planets, in agreement with a well-known Aristotelian dictum. In later work, Kepler kept the main assumptions of this cosmology. In the *Astronomia nova* (1609) he tried to develop a new physics, bringing together explanations from material causes and rigorous mathematical demonstrations.[19] Apart from this, he kept his belief in universal harmony, and worked hard to reconcile his geometrical hypotheses with Brahe's precise observations. The most relevant result of this effort was his *Harmonices Mundi*, published in 1619.

Kepler's Epistemology

According to Kepler, astronomy cannot do without cosmological assumptions. Thus, he opposed the mathematic-conventionalist interpretation of Copernicus's work (generally referred to by historians of science as the "Wittenberg interpretation").[20] At that time, most European mathematicians shared the conventional point of view, first endorsed by German scholars, that restricted astronomical theory to its predictive capability, regardless of its physical tenability. In order to defend against the apparent inconsequentiality of this approach, they alleged the logical assumption (*necessitas syllogistica*) that false hypotheses might lead to true conclusions. The theologian Osiander had already presented this argument in the anonymous preface of the first edition of Copernicus's *De revolutionibus*, in order to preserve "mathematical" astronomy from conflict with peripatetic physics and Biblical exegesis. Legitimized by conventionalism, many post-Copernican mathematicians (e.g. Wittich from Breslau, and the Scotsman Liddel, who taught in Rostock and Helmstedt) demonstrated the geometrical equivalence of different planetary models, regardless of their physical reality.[21]

---

[18]   See A. Segonds, *Introduction*, in J. Kepler, *Le Secret du Monde [Mysterium cosmographicum]* (Paris, 1984), p. xxvi: "so one of the clearest results of *Mysterium cosmographicum* was inducing Mästlin to abandon his prudence."

[19]   E. Rosen, "Kepler's Place in the History of Science," in Beer and Beer, *Kepler*, pp. 279–85; p. 280.

[20]   Westman, "The Melanchthon Circle."

[21]   See C.J. Schofield, *Tychonic and Semi-Tychonic World Systems* (New York, 1981); and Gingerich and Westman, "The Wittich Connection."

In *Astronomia nova* Kepler drew anti-conventionalist arguments from the French philosopher Petrus Ramus.[22] Yet he did not share Ramus's general idea of freeing astronomy from all hypotheses. Besides Ramus, the Italian Neoplatonic philosopher Francesco Patrizi also maintained that geometrical hypotheses are useless, because celestial bodies move irregularly, and only God's Providence accounts for their motions. In the first chapter of *Apologia Tychonis*, Kepler refuted Patrizi's cosmology. He accepted both the free motion of planets in space (*"Primum hoc illi [Patricii] facile concessero, solidos orbes nullos esse"*), and intelligent universal design (*"Neque nego, planetarum circuitus ratione summa administrari"*),[23] but considered hypotheses necessary, because he held that God's creation realizes uniform and perfectly circular motions (*"ut uniformem et quam fieri potest regolarissimum circulum describant"*).[24] Kepler was in disagreement with the Italian philosopher also on another cosmological issue: the infinite space beyond the fixed stars. In fact, according to Kepler, geometrical perfection implied cosmological proportion and finiteness.

In *Apologia Tychonis*, he went deeply into the epistemological status of hypotheses. In the section *Quid sit hypothesis astronomica*, he criticized the imperial mathematician Ursus for embracing conventionalism. Against such a "vain" approach to astronomy, Kepler affirmed astronomical hypotheses to be both useful and true. In fact, false hypotheses, like lies, generate innumerable errors, especially in physics. However, Kepler distinguishes various meanings of "hypothesis." First of all, in geometry, it means the starting point of a demonstration (*"certum quodam initium"*), similar to the foundations of a building in architecture (*"fundamenta domus"*). Secondly, in Aristotle's logic, "hypothesis" means the premise of a syllogism. Thirdly, in astronomy, there are two meanings of "hypothesis." In origin, this term referred to empirical data, on which theory is based. The meaning of "general conception" (*"summam quodam conceptionum celebris alicuius artificis, ex quibus totam ille rationem motuum coelestium demonstrat"*)[25] became then usual. According to Kepler, conventionalism brings out an incorrect analogy between hypotheses in astronomy, *suppositiones* in geometry (where different presuppositions can demonstrate the same thesis), and premises in logic (where it is true that F→T).

Thus conventionalism entails an equivocation: the confusion of geometrical, logical, and astronomical hypotheses. Unlike geometrical models, astronomical systems cannot be equivalent, because they imply different physical consequences (*"Nam si in geometricis duabus hypothesibus conclusiones coincidant, in*

---

[22]   Kepler, *Astr. nova*, p. 6; and W.T. Danahue (trans.), *New Astronomy* (Cambridge, 1992), p. 28.

[23]   J. Kepler, *Apologia Tychonis contra Ursum*, in *Opera omnia*, ed. C. Frisch (Frankfurt am Main and Erlangen, 1858), p. 247.

[24]   *Ibid.*

[25]   *Ibid.*, p. 239.

physicis tamen qualibet habebit suam peculiarem appendicem").[26] For instance, a Ptolemaic astronomer like Magini, though using Copernicus's tables, must consider the parallax of Mars greater than the sun's—an incorrect consequence of geocentrism.[27] Another example is taken from a comparison between Copernicus and Brahe. In the absence of an observable stellar parallax, the distance of the stars is much greater according to heliocentrism. According to this theory, in fact, their distance is deduced from the diameter of the earth's orbit around the sun, instead of the earth's diameter only, as is the case with geocentrism and geo-heliocentrism. As a matter of fact, Kepler remarked that all cosmological hypotheses bear important and problematic consequences, such as the earth's motion and the immensity of the sky according to Copernicus; the troublesome rotation of the planets around a rotating sun according to Brahe; and the uttermost rapidity of daily stellar motion according to Ptolemy (and Brahe).

Kepler encouraged astronomers to discover the "true" motions, which agree with theory (*"vias vero veras invenire, opus esse astronomiae contemplative"*).[28] It was his belief that hypotheses must be "true in every respect" (*"et proinde hypothesibus hoc est proprium ... ut sint undiquaque verae"*), as was the case with Copernicus, *pauculis mutatis.*[29]

Benedetti's Mathematical Philosophy

Perhaps one way to sum up Kepler's position would be to say that he held that astronomy ought to be considered part of natural philosophy—a discipline on which peripatetic philosophers claimed the monopoly. In *Apologia Tychonis*, he wrote explicitly: "The astronomer ought not to be excluded from the community of philosophers who inquire into the nature of things."[30] A similar opinion was expressed by Giovanni Battista Benedetti, who was mathematician to the Court of Savoy in Turin, and was regarded by Kepler as one of the few Italians who were not "asleep."[31]

In *Diversarum speculationum mathematicarum et physicarum liber* (Turin, 1585), Benedetti asserted that mathematicians can legitimately deal with natural

---

[26]   *Ibid.*, p. 240.

[27]   Kepler's remark on Magini is incorrect, as the Italian astronomer endorsed a particular version of geo-heliocentrism, illustrated in G.A. Magini, *Novae coelestium orbium theoricae* (Venice, 1589).

[28]   *Apol. Tycho*, p. 248.

[29]   *Ibid.*, p. 241.

[30]   N. Jardine (ed. and trans.), *The Birth of History and Philosophy of Science: Kepler's "A Defence of Tycho against Ursus"* (Cambridge, 1984), p. 144.

[31]   J. Kepler, letter to S. Hafenreffer (Prague, 1606), in *Gesammelte Werke* (Munich, 1951), p. 390: *"Itali somniant (praeter unum Grumandinum et Ioh. Baptistam Benedictum, Clavius enim Germanus est)."*

issues. His purpose was, indeed, to mathematize physics. In a letter to the Venetian patrician Domenico Pisani, included with the title *De philosophia mathematica* in the already mentioned *Diversarum speculationum liber*, Benedetti reaffirmed the philosophical status of his discipline, at the same rank as physics, metaphysics, and morals, by reason of the certainty of its demonstrations ("*certitudo suarum conclusionum*").[32] Like his teacher Niccolò Tartaglia and his correspondent Pietro Catena, a professor at the University of Padua,[33] Benedetti opposed Alessandro Piccolomini's and Benedict Pereira's Aristotelian refutation of the possibility of explaining nature by the means of mathematics.[34] As a direct consequence of this epistemology, he dismissed the traditional distinction between physics and mathematics also in cosmology. That is, he refused to divide the investigation of "causes" from that of calculation.[35] This anti-conventionalist opinion was combined with adherence to the Copernican system.[36]

Benedetti corresponded with Patrizi,[37] agreeing with him that space is infinite above the fixed stars, with the difference that he conceived of the planetary system as heliocentric. In a letter to the Savoian court historian Pingone, Benedetti maintained that space is boundless: "it is not necessary that the place of fixed stars be terminated by any convex-concave surface."[38] Benedetti's argument is aprioristic, in that he deduces the reality of infinite space from its mere possibility. In fact, he appears to believe that, once the rational possibility is ascertained, it is not necessary to demonstrate that the universe is infinite, but rather that it is limited. Moreover, according to Benedetti, the sky is fluid, so the earth and the other planets move in a motionless *aer*.

The title of another letter, *De ... infinito spacio extra coelum, coelique figura*, points out the distinction between infinite space (*spacium*) and finite heaven

---

[32]   G.B. Benedetti, *Diversarum speculationum mathematicarum et physicarum liber* (Turin, 1585), *De philosophia mathematica*, p. 298: "*Miror quod cum in Aristotele sis versatus, in tuis tamen scriptis philosophum a Mathematico separes, quasi mathematicus non sit adeo philosophus, ut est naturalis, et metaphysicus, cum multo magis quam ii philosophus sit appellandus, si ad veritatem suarum conclusionum respiciamus.*"

[33]   *Ibid.*, p. 371.

[34]   See A. De Pace, *Le matematiche e il mondo. Ricerche su un dibattito in Italia nella seconda metà del Cinquecento* (Milan, 1993), pp. 228–9.

[35]   See M. Di Bono, *Le sfere omocentriche di Giovan Battista Amico nell'astronomia del Cinquecento* (Genoa, 1990); and M.A. Granada and D. Tessicini, "Copernicus and Fracastoro: The Dedicatory Letters to Pope Paul III, the History of Astronomy, and the Quest for Patronage," *Studies in History and Philosophy of Science*, 36 (2005): pp. 431–76.

[36]   M. Di Bono, "L'astronomia copernicana nell'opera di Giovan Battista Benedetti," in *Cultura, scienze e tecniche nella Venezia del Cinquecento* (Venice, 1987), pp. 283–300.

[37]   G. Claretta, "Lettere tre di Francesco Patrici a Giambattista Benedetti matematico del Duca di Savoia," in *Miscellanea di Storia Italiana* (Turin, 1862), vol. 1, pp. 380–83.

[38]   Benedetti, *Div. spec.*, p. 256: "*nulla [est] necessitas, ut locus fixarum terminaretur aliquibus superficiebus, convexa scilicet, et devexa.*"

(*coelum*). The world, plunged in cosmological infinity, is spherical by reason of an aprioristic consideration of the "economy" of this solid figure, "because no body can be defined by a limit more briefly than by a sphere."[39] Yet Benedetti's Pythagorean philosophy is remarkably different from Kepler's. Benedetti was convinced, unlike Kepler, that there is an ontological hiatus between ideal perfection and its concrete realization. He kept closer to Proclus's doctrine, reflected in Renaissance Neoplatonist theory, of the ontological-epistemological *medietas* of mathematical beings, placed between imperfection and perfection, sensible and intelligible reality—or better, between creation and God's mind.

Benedetti worked on calendar reform and ephemerides long enough to notice that there were no absolutely reliable predictions, but only more-or-less exact calculations and tables. Nonetheless, he wrote an apology for ephemerides, *Defensio ephemeridum*, in which he defended the validity of astronomical calculations in spite of their intrinsic limits.[40] He tended to explain the incomplete regularity of celestial motions through the Platonic doctrine of natural imperfection, in a way similar to cardinal Cusanus, who, in the fifteenth century, argued that the only absolute equality is that of God with himself ("*praecisam aequalitatem solum deo convenire*").[41]

Benedetti shared Kepler's aprioristic assumption of celestial harmony. The Italian mathematician reflected thereupon in a section of his *Diversarum speculationum*, in which he refuted peripatetic physics, *Disputationes de quibusdam placitis Arist[otelis]*. In chapter XXXIII, *Pythagoreorum opinionem de sonitu corporum coelestium non fuisse ab Aristotele sublatam*, Benedetti denied that the "sound of celestial bodies" is the material production of any sounds. Unlike Kepler, he thought there was no harmonic proportion among planetary motions, as there was no perfect astronomical geometry. Rather, he reduced the Pythagorean doctrine of world harmony to Divine Providence.[42]

It is remarkable how different are the conclusions that Benedetti and Kepler reached from the same philosophical starting-point. They were both convinced of the geometrical-rational structure of the universe first taught by Pythagoras; but the former believed that Divine Providence must manifest itself through infinite space, whereas, according to the latter, God's creation must be finite, harmonic, and proportional. Furthermore, Benedetti did not suppose that nature can fully realize mathematical perfection, whereas Kepler held the opposite view.

---

[39]   *Ibid.*: "*quia nullum corpus a breviori termino quam a spherico terminari potest.*"

[40]   *Ibid.*, pp. 228–48.

[41]   Nicolaus de Cusa, *De docta ignorantia/Die belehrte Unwissenheit* (Hamburg, 1999), vol. 2, p. 4.

[42]   Benedetti, *Div. spec.*, p. 191: "*Quod autem attinet ad motus, ad magnitudines, ad distantias et ad influxus, nihil est, quod hisce proportionibus conveniat, sed quia haec omnia dependent ab infinita et divina providentia Dei, necessario sit ut istae velocitates, eae magnitudines, distantiae et influxus, talem ordinem et respectum inter se ipsa et universo habeant, qualis perfectissimus sit.*"

Bruno's Infinite and Homogeneous Universe

Benedetti was rather elusive about the metaphysical reasons for cosmological infinity. In contrast, his contemporary Giordano Bruno was explicit and exhaustive on this issue.[43] In the Italian dialogue *De l'infinito, universo e mondi* (1584) and in the Latin poem *De immenso et innumerabilibus* (1591), Bruno analyzed, discussed, and refuted Aristotle's arguments against the infinite universe. Moreover, in the eighth book of *De immenso*, he criticized the infinite cosmology of Palingenius's *Zodiacus Vitae* (1534). He mocked this author as one of those "sleepers with the rabble who realize that they are dreaming, and try to shake off their sleep; but presently they dream that they are awake, having merely changed the vision of sleep, not driven it away."[44] Palingenius, like the later Patrizi and Benedetti, believed space to be infinite, but the world finite, surrounded by incorporeal light. In contrast, Bruno stated that God cannot create a finite world and a heterogeneous universe, as a consequence of his wisdom, power, love, supremacy, glory, and life.[45] A God of infinite power but finite creative action would be infinitely "jealous" and finitely good. There follows Bruno's infinite and homogeneous cosmology, as an application of the so-called "principle of plenitude."[46] One of the main sources of Bruno's cosmology is cardinal Cusanus's *De docta ignorantia*, from which he derived arguments for the infinite worldly "sphere," as well as for the earth's motion. It is notable that Bruno often coupled the names of Cusanus and Copernicus.[47]

Bruno distinguishes the "infinite sphere" of the world from the "finite spheres" of celestial bodies. These are either "fires"/suns, or "waters"/earths.[48] According

---

[43]   See P. Omodeo, "La Stravagantographia di un 'filosofo stravagante'," *Bruniana & Campanelliana*, 14/1 (2008): pp. 11–23; and "La cosmologia infinitistica di Giovanni Battista Benedetti," *Bruniana & Campanelliana*, 16/1 (2009): pp. 181–90.

[44]   G. Bruno, *De immenso et innumerabilibus*, VIII,2, in *Opera latine conscripta* (Stuttgart-Bad Cannstatt, 1962), vol. 1, 2, p. 292: "*Quidam somniantes cum vulgo, somniare se intelligunt, conantur somnum excutere, sed mox se vigilare somniant, somni mutata specie, non abacta.*"

[45]   *Ibid.*, p. 293.

[46]   See A.O. Lovejoy, *The Great Chain of Being* (Cambridge, MA, 1936); A. Koyré, *From the Closed World to the Infinite Universe* (Baltimore and Oxford, 1957); and M.A. Granada, "Il rifiuto della distinzione tra potentia absoluta e potentia ordinata di Dio e l'affermazione dell'universo infinito in Giordano Bruno," *Rivista di storia della filosofia*, 49/3 (1994): pp. 495–532.

[47]   See for instance Bruno, *De imm.*, III,9, vol. 1,1, p. 381: "*Mirum, o Copernice, ut e tanta nostri seculi caecitate quando omnis philosophiae lux cum ea quae aliarum quoque rerum inde consequentium est, extincta jacet, emergere potueris; ut ea quae suppressiore voce proxime praecedente aetate in libro De docta ignorantia Nicolaus Cusanus enunciabat, aliquando proferres audacious.*"

[48]   G. Bruno, *Articuli adversus mathematicos*, in *Opera latine conscripta*, vol. 1, 3, art. 134, p. 72: "*Nobis sphaera universalis est unum continuum universum infinitum immobile,*"

to him not only is the universe infinite, but it also includes innumerable worlds, or better, heliocentric planetary systems (*"synodis ex mundis"*).[49]

The difference between Bruno's conception and those of Kepler and Benedetti is noteworthy. It emerges clearly from his *Articuli ... adversus huius tempestatis mathematicos atque philosophos* (Prague, 1588), article 142, which rejects every possible limitation of the *coelum*: "For us, neither similitude to archetypes, nor convenience of capacity, nor necessity of distinction, nor the impossibility of the vacuum, nor the unsuitability of the penetration of bodies means that the heavens are spherical."[50]

Nature and Mathematics in Bruno

In natural philosophy, Bruno always preferred physical to mathematical explanation. Notably, he criticized Copernicus's "excess" of mathematics, at the expense of philosophical and cosmological speculation. In his conception, the universe is similar to an immense animal, living in every part.[51] A vital impulse (*"vis animalis"*) permits planetary motions (*"omnium principium motus intrinsecus est animalis appulsus, atque spiritus universum exagitans"*).[52] Planets have, in fact, a sensible and intellectual soul in order to perform their functions. Universal vitalism implies, according to Bruno, the becoming of all beings, called *vicissitudo* or, in Italian, *vicissitudine*.[53] This doctrine is closely related to atomism. Atoms are what

---

seu in quo consistentia sunt numero infinitae sphaerae seu particulares mundi. Haec astra sunt alia quidem ignea, scintillantia, quae videntur fixa, soles, maresque dii, alea vero aquea, circa hos mobilia, puta telluresm deaeque antiquis."

[49]   See M.A. Granada, "Synodis ex mundis," *Bruniana & Campanelliana*, 13/1 (2007): pp. 149–56.

[50]   Bruno, *Articuli*, art. 142, p. 73: *"Coelum neque similitudo archetypi, neque commoditas ad capiendum, neque distinctionis necessitas, neque vacui impossibilitas, neque penetrantium corporum inconvenientia faciet nobis esse rotundum."*

[51]   See Bruno, *De l'infinito universo e mondi*, in *Dialoghi filosofici italiani* (Milano, 2000), p. 373: *"Oltre dico, che questo infinito et immenso è uno animale, benché non abia determinata figura, e senso che si referisca a cose esteriori: perché lui ha tutta l'anima in sé, e tutto lo animato comprende, et è tutto quello."*

[52]   Bruno, *Articuli*, art. 141, p. 72; and see Bruno, *La cena de le Ceneri*, in *Dialoghi filosofici italiani* (Milan, 2000), p. 80: *"Gli astri [...] come danno la vita e nutrimento alle cose, [...] cossì e molto maggiormente hanno la vita in sé: per la quale, con una ordinata e natural volontà, da intrinseco principio se muoveno alle cose o per gli spacii convenienti ad essi."* Cf. Engl. transl. by S.L. Jaki, *The Ash Wednesday Supper* (Paris, 1975), p. 115: "Since they [the celestial bodies] give life and nourishment to the things ..., they have in themselves life more abundantly, by which, as if by a directed and natural will [stemming] from an intrinsic principle, they tend toward other things and through spaces which are convenient to them."

[53]   M. Ciliberto, *La ruota del tempo* (Rome, 1987).

persist in natural transformations: the ephemeral life of compounds is caused by the movement of their never-perishing constituents. Furthermore, Bruno reduces all changes to local (atomic) motion, an idea explicitly drawn from Democritus, Epicurus, and Lucretius.[54]

In *Articuli adversus mathematicos*, and above all in *De triplici minimo et mensura* (1591), Bruno regards mathematics as a doctrine dealing with bodily properties. Thus, the finite divisibility of the bodies (atoms are, indeed, the last indivisible components of matter) prohibits dividing numbers to infinity. Irrational numbers are banished. For the same reason, Bruno also denies the possibility of squaring the circle, and regards geometrical figures as irreducible and qualitatively different, "because a polygonal figure and a circular one cannot be composed of the same number of parts."[55] It is not possible to compare "either the square with the circle, nor the square with the pentagon, nor the triangle with the square, nor a figure of a species with a figure of another species."[56] Moreover, since geometrical figures reveal the properties of atomic compounds, Bruno is convinced that reflection on geometrical figures, and on numerical sequences expressing their augmentation, leads to comprehension of bodily basilar characteristics. Hence the complex numerology he expounds in *De triplici minimo*. In his opinion, the possibility of a mathematical natural science depends on the atomic structure of reality. Its validity is strictly related to Democritean physics. Bruno writes that scholars of geometry violate the laws of nature, if they maintain infinite numerical divisibility, because they neglect atomism: "Thus, the surveyor who divides to infinity what has a precise quantity is wrong, does not follow nature [*naturae vestigia*], and never grasps it, nor agrees with it in any respect."[57]

The minimum is only intelligible, that is, it is conceivable but not given to the senses. In both physics and mathematics the five senses are therefore useless. On this basis, Bruno distinguishes primary (true) properties from secondary (apparent) properties of natural beings: "With the eyes we perceive light, colour and motion, but with them we cannot grasp the truth of that colour and that light which we perceive, nor distinguish it from appearances of the same species."[58] Furthermore,

---

[54]   Bruno, *La cena de le Ceneri*, p. 82.

[55]   Bruno, *De triplici minimo et mensura*, II,8, in *Opere latine conscripta*, vol. 1, 3, p. 215: "*veluti laterale nequit sphaeraleque tantis atque tot esse tomis.*"

[56]   *Ibid.*, p. 217: "*Iam cum in natura sit definitum minimum, neque secundum actum neque secundum rationem circulo quadratum, immo nec quadratum pentagono, neque triangulum quadrato, neque tandem figurae ullam speciem cum alterius speciei figurae possibile est aequare.*"

[57]   *Ibid.*, 1, 7, pp. 154–5: "*Ergo errat mensor certum sine fine resolvens quantum, naturae nusquam vestigia lustrans, nusquam illa attingens, non ullis sortibus aequans* [...] *quoties per inane vagatur.*"

[58]   *Ibid.*, 2, 3, p. 194: "*Oculo enim lucem, colorem atque motum videmus, verum autem oculo videre non possumus; neque etenim in oculo vis ea sita est, qua hunc esse verum colorem lucemque veramm diiudicemus, et ab apparentibus eiusmodi distinguamus.*"

the becoming of all compounds implies that macroscopic phenomena (such as planetary motions) are not perfectly mathematical. Their measurement is extremely difficult, as absolute exactness is revealed to be impossible:

> Nam rerum numeros alios momenta reportant
> Singula, quae celeri nulla virtute coirent;
> Organa qui poterit reputari exacta dedisse
> Heic ubi nec fluxus eadem est dimensio, ut inde
> Terminus a reliquo aeque absistat, vel semel unus?

> [Singular aspects of things, which slowly change, express different numbers; who can hold that instruments provide accurate measuring, since the becoming of things does not keep a constant rhythm and no limit remains at the same distance from another?][59]

This consideration does not lead to skepticism. Bruno explains that, for practical purposes such as measuring, one should be satisfied with approximation (*"aequale magis suscipere atque minus"*).[60] Approximation is necessary in the quantification of all natural phenomena, including the astronomical.[61]

To sum up, Bruno's view of mathematics and mathematical explanation directly derives from his vitalist and atomist natural philosophy. In particular, his atomistic foundation diverges from the Platonic (and Proclean) conception. According to that tradition, in fact, the mathematical order of nature relies on ideal geometrical entities, and on an ontological hierarchy of reality. This viewpoint was shared by Kepler and Benedetti, though they were in disagreement as to the degree of mathematical perfection of the material world. In contrast, Bruno's mathematics is based on the so-called "minimum," the mathematical indivisible point, counterpart of the physical atom.

Kepler–Galilei–Bruno

About 1610 Kepler seriously confronted Bruno's cosmology and philosophical approach. It happened after the publication of Galilei's *Sydereus Nuncius* (1610), which described new telescopic discoveries: the mountainous surface of the moon, a large number of newly observed stars and, above all, some satellites of Jupiter. Galilei had already declared his adherence to the Copernican system in the dedication to Cosimo De' Medici, suggesting that the discovery of the

---

[59]   *Ibid.*, 2, 5, p. 203.

[60]   *Ibid.*, 5, 2, p. 303.

[61]   A. Bönker-Vallon, "Bruno e Proclo: connessioni e differenze tra la matematica neoplatonica e quella bruniana," in Eugenio Canone (ed.), *La filosofia di Giordano Bruno. Problemi ermeneutici e storiografici* (Florence, 2003), pp. 129–44.

Jovian satellites proved that the moon was not the only planetary satellite of the heliocentric system.[62] In fact, the exception of the moon had been regarded as an inconvenient aspect of Copernicus's theory.

Kepler was informed of Galilei's discoveries by his friend Wackher von Wackenfels, who was an imperial functionary in Prague. Wackenfels, an adherent of Bruno's natural philosophy, believed that the recently-discovered planets could prove cosmological infinity. As he was not well informed, he thought that the new planets rotated around some fixed star, whose cosmological necessity had been maintained by Bruno.[63] Furthermore, Bruno's cosmology was endorsed also by another acquaintance of Kepler's, the English mathematician and botanist Edmund Bruce, who, between 1599 and 1605, was in epistolary communication with him from Italy, acting as an intermediary between him and Galilei.[64]

Kepler's initial concern about the new planets was that this discovery could invalidate the cosmic order presented in the *Mysterium*. As an answer to Galilei he wrote *Dissertatio cum Nuncio Sydereo* (1610), where he declared his satisfaction that the new planets rotated around a planet, and not around a star: "If you had discovered any planets revolving around one of the fixed stars, there would now be waiting for me chains and a prison amid Bruno's innumerabilities, I should rather say, exile to his infinite space."[65] As it was, Kepler rejoiced at Galilei's telescopic observations, agreeing with him that his discoveries reinforced the heliocentric system, showing other satellites besides the moon.

In the *Dissertatio* Kepler also meant, however, to give Galilei a lesson on epistemology. Sustaining the superiority of an aprioristic approach, he did not hide his conviction of the pre-eminence not only of his work, but also of Bruno's, towards Galilei's more empirical astronomy. He celebrated, among the ancients, Pythagoras, Plato, and Euclid, who were guided by the light of their reason only, and held that the universe respects a divine and geometrical law of harmony. Copernicus, revealing the true planetary motions and the centrality of the sun, established "mere" facts, whereas Kepler took credit for the discovery of the "secret causes" of the world's construction:

> For the Glory of the Creator [*Architectus*] of this world is greater than that of the
> student of the world, however ingenious. The former brought forth the structural
> design from within himself, whereas the latter, despite strenuous efforts,

---

[62]   G. Galilei, *Sidereus Nuncius*, in *Le Opere*, *Edizione Nazionale* (Florence, 1930), vol. 3, 1, p. 56, 95.

[63]   J. Kepler, *Dissertatio cum Nuncio Sidereo*, in *Gesammelte Werke* (Munich, 1941), vol. 4, p. 289. See M.A. Granada, "Kepler and Bruno on the Infinity of the Universe and of Solar Systems," *Journal for the History of Astronomy*, 39 (2008): pp. 469–95.

[64]   See M. Bucciantini, *Galileo e Keplero. Filosofia, cosmologia e teologia nell'Età della Controriforma* (Turin, 2003), pp. 93–116.

[65]   E. Rosen (trans.), *Kepler's Conversation with Galileo's Sidereal Messenger* (New York and London, 1965), pp. 36–7.

scarcely perceives the plan embodied in the structure. Surely those thinkers who intellectually grasp the causes of phenomena, before these are revealed to the senses, resemble the Creator more closely than others who speculate about the causes after the phenomena have been seen.[66]

Kepler firmly believed that *good science is* a priori. On this matter he clearly agreed with Bruno and Benedetti. Moreover, insofar as Kepler committed himself, and subsequent astronomy, to the discovery of hidden laws—the secrets of the universe—he did so precisely on this *a priori* basis.

Conclusion and Future Research

In the late Renaissance, as we have seen, the most influential post-Copernican astronomers founded their views on metaphysical speculations about the perfection of the universe, though differently conceived. Despite the fact that empirical astronomy—developed primarily by Brahe—permitted better predictions, the *a posteriori* approach began to be regarded as less important than rational cosmology. This is clearly revealed, for instance, by Kepler's judgment on his predecessors and Galilei. Notably, the emerging cosmological debate was not so much a criticism of empirical astronomy as an internal confrontation among concurring aprioristic epistemologies and philosophies of nature.

Despite their common background—a Pythagorean and Platonic anti-Aristotelian philosophy, implying cosmic perfection and universal harmony—Kepler, Benedetti, and Bruno proposed very different philosophical foundations for nature and astronomy: a strictly geometrical science (Kepler); a more moderate, slightly skeptical, mathematical one (Benedetti); and a vitalist, infinitely multiform, and anti-geometrical view of nature (Bruno). The differences among these three thinkers reveal the importance of epistemological and aprioristic (theological, natural, metaphysical, and cosmological) reflections for the development of the modern scientific world view. According to them, indeed, the defense of the new science could not be separated from strong ontological and epistemological convictions.

Besides, their disagreement on cosmology, perfection of the world and mathematics reveals that, between the sixteenth and the seventeenth centuries, there was no standard Copernicanism, despite the simplified image, later promoted by Galilei, of a struggle between two chief world systems, Copernican and Ptolemaic. In fact, the concurring cosmologies were more than two, even within the heliocentric framework. Copernicus's work did not lead to a shared conception, or to a unified natural science, but rather to a plurality of possible "Copernicanisms." Instead of a standard view of nature and method, early-modern

---

[66] *Ibid.*, p. 38.

science gave birth to a pluralistic and dubitative attitude, expressed through an intense and eclectic philosophical debate.

Chapter 7
Discovery in *The World*:
The Case of Descartes

Jacqueline Wernimont

Asked to characterize Cartesian discovery, one might immediately turn to Descartes's works on mechanics, optics, mathematics, or even metaphysics. In each of these cases, the object of discovery would be something actual, something that, theoretically at least, obtained in Descartes's early-modern world. While the Cartesian mind and body might challenge how we use the word "actual," depending on our readings, Descartes is nevertheless always writing about how actual human entities work. Except for when he's not. *Le Monde*, or *The World*, which became available only posthumously, differs from much of Descartes's other work in that it uses a self-consciously imaginative narrative for philosophical inquiry.[1] While this experiment in creative writing subsequently shaped its author's use of the "fable" in the *Discourse on Method, The World* is the only one of Descartes's texts to be structured by the logic of a literary possible-worlds narrative. It is the text that demonstrates most clearly Descartes's production of a kind of knowledge that we might call poetic.

Descartes's mechanist "paper-world" has a moon, a sun, and is surrounded by other planets and stars; its seas ebb and flow; its light refracts; it is inhabited by men that have souls, bodies, desires, and drives.[2] Rather than describing discoveries about the actual world, this text offers insight into a *possible* world, one where nature operates like a machine, obviating the need for God's creation of man, animals, and so on. In explicitly engaging with a possible world, *The World* is different from early-modern mimetic literary texts that reproduce, in one form or another, the actual world. Further, unlike utopian reconstructions of the real, such as Thomas More's *Utopia* or Francis Bacon's *New Atlantis, The World*

---

[1] René Descartes, *The World and Other Writings*, ed. Stephen Gaukroger (Cambridge, 1998).

[2] Neil Rhodes and Jonathan Sawday open their collection, *The Computer in the Renaissance*, with a description of the post-Gutenberg era as itself a "paperworld ... a place of the imagination and intellect rather than a geography of curious beasts, peoples, and plants." While they were describing a historical moment in the actual world, I find their description equally apt for Descartes's *The World*. See Rhodes and Sawday (eds), *The Computer in the Renaissance* (New York, 2000), p. 1.

emphasizes *novel* creation alongside the unique discovery enabled by reading imaginative texts.[3]

While Descartes had precedents for his possible world (Lucian of Samosata wrote a *True History* of a possible world in the moon in the second century A.D.), *The World* is among the earliest examples of explicit possible-world literary creation for epistemological purposes. Consequently, it offers a surprising perspective on what "discovery" entailed for the early-modern. While seventeenth-century Europeans were enchanted by empirical and ocular demonstration, *The World* attests to the power of imaginative stipulation for discovery, and to the ways in which discovery occurs in and through language. Cartesian mechanism is a textual event first, one that is actively called into being by the text. While the text refers to a "non-existent" possible world, it nevertheless retains the epistemological status of the "real." Only later, through processes of analogy and comparative analysis, does the object of readerly and writerly discovery return to the actual world.

Descartes recognized that his new mechanist universe was dangerous in the intellectual climate of the 1630s, and he chose to suppress the text rather than have it and himself subject to the scrutiny of the Catholic Church.[4] While he published some of the mechanical arguments from *The World* and made reference to the text often in other publications such as the *Discourse* (1637) and the *Principles of Philosophy* (1644), the complete text of *The World* is lost to us. What we do have are portions of the text posthumously published by Claude Clerselier as *The World, or the Treatise on Light*, and the *Treatise on Man* (1664). The Clerselier *World* is fragmented. The two treatises suggest that intervening text may have existed or was planned (the *Treatise on Man* opens by referring to text which is not extant), and they suggest that a third treatise on the soul was planned, though it is unclear if this text was ever written.[5] Nevertheless, the *Treatise on Light* and the *Treatise on Man* preserve the structure and much of the content of a remarkable text that is simultaneously Cartesian natural philosophy and, in modern generic terms, a "possible-worlds fiction."[6]

---

[3]   Joaquin Martinez Lorente argues for a limited application of possible-worlds theory to More's *Utopia*, which preserves it as an "Ideal World" text. Lorente's reading differs from my own by subsuming the ideal or utopian under the possible. See Joaquin Martinez Lorente, "Possible World Theories and the Two Fictional Worlds of More's Utopia: How Much (and How) Can We Apply," *Sederi*, 6 (1996): pp. 117–23. Available at: sederi.org/docs/yearbooks/06/6_13_martinez.pdf.

[4]   René Descartes, *The Philosophical Writings of Descartes*, eds John Cottingham, Robert Stoothoff, and Dugald Murdoch (3 vols, Cambridge, 1991), vol. 3, p. 28. This edition hereafter cited as CSM.

[5]   Descartes, *The World*, p. vii.

[6]   On possible worlds see Lubomir Doležel, *Heterocosmica: Fiction and Possible Worlds* (Baltimore, 1998); Ruth Ronen, *Possible Worlds in Literary Theory* (Cambridge, 1994); Thomas L. Martin, *Poeisis and Possible Worlds: A Study in Modality and Literary*

To date, literary theorists have used possible-worlds theory as a way to develop theories of fictionality.[7] But for Descartes, to say something was "fictitious" was to disparage its intellectual value and epistemological rigor; consequently this essay only glances at the issue of "fiction." Instead, possible-worlds theory helps us to discriminate between writing that operates in a conditional or propositional vein and the traditional early-modern mimetic modes of writing. Insofar as *The World* demonstrates that early-modern "discovery" could happen between the covers of a book—that discovery need not be predicated on the real, but could occur through simultaneously aesthetic and pragmatic channels—it has a great deal to offer modern theory.

Poetic Knowledge

Matthew Jones observes that Descartes held a theory of poetic knowledge, strongly influenced by Jesuit training in rhetoric.[8] However, neither classical rhetoric nor the Jesuit appropriation of it included the creation of possible worlds in the arts of persuasion. Instead, rhetoric remained largely mimetic in one form or another. Consequently, Jones's excellent work on Descartes's uses of his rhetorical training can be extended if we recognize Descartes's innovations, *vis-à-vis* that training, as well. Descartes utilized possible scenarios and possible worlds to enable the kind of *enargia*, or vivid lively description, valued by Jesuit rhetoricians and central to knowledge according to Descartes. The constraints of creating a coherent possible world required that Descartes simplify the extremely complex issue of creation (down to the principle of motion), allowing clear articulation of the linkages between elements. Insofar as he was able (in his own estimation) to make those linkages and the primary elements of his new world as clear as other kinds of intuitions, Descartes felt that he was able to offer a fable as epistemologically valuable as any observation or representation of the "real" world. Where traditional rhetoric used language to represent the world mimetically, Descartes extended that tradition by including new worlds, thereby putting the fable to rigorous philosophical work.

Cartesian poetic knowledge is characterized by knowledge of "unifying causal structures," and it is in order to present such unifying structures—despite the logical contradictions and theological implications that such structures would have for the *actual* world—that Descartes turns to the literary production of *possible* worlds.[9] While relatively little critical attention has been paid to Descartes's use of

---

Theory (Toronto, 2004); and Mary Baine Campbell, *Wonder and Science: Imagining Worlds in Early-Modern Europe* (Ithaca, 1999).

[7] See Ronen; Martin; Doležel; and Umberto Eco, "Small Worlds," *Versus: Quaderni di Studi Semitotici*, 52/3 (1989): pp. 53–70.

[8] Matthew Jones, *The Good Life in the Scientific Revolution* (Chicago, 2006), p. 28.

[9] *Ibid.*, pp. 58–70.

poetic modes, an interest in the use of poetic language for philosophical inquiry is evident in his early work. In his notebooks he observes:

> it may seem surprising to find weighty judgments in the writings of the poets rather than the philosophers. The reason is that the poets were driven to write by enthusiasm and the force of imagination. We have within us the sparks of knowledge, as in a flint: philosophers extract them through reason, but poets force them out through the sharp blows of the imagination, so that they may shine more brightly.[10]

Poets alone utilize a creative force that, upon impact with the intellect, produces intuitions so bright that they compel belief in their certainty. Thus poetic writing offers superior access to knowledge. Imagination is a critical tool for Cartesian knowledge; in the *Rules for the Direction of the Mind* (1628), Descartes suggests that in order for the intellect to perceive truth it must "be assisted by the imagination."[11] The "sparks of knowledge" shine in the hands of poets, whereas philosophical "extraction" offers the reader something considerably more dull.

Mimeticism and Possible Worlds

It is important to make a clear distinction between *mimetic* and *possible* literary worlds. *Possible worlds* differ from other literary productions by rejecting a mode of signification that depends on reference to an actual world. While possible-world theory has become increasingly popular for modern philosophers and literary critics flummoxed by fictional and impossible language, the issue of the nature of literary signification was endemic to early-modern Europe. Sixteenth- and seventeenth-century mimetic theories generally operated along either the Aristotelian or Platonic axis. For Aristotle, language faithfully *simulates* phenomena, and is firmly grounded in experience of the "actual." The poet's or dramatist's function is to "express the universal," imitating not just what has happened, but the "sort of thing that a certain kind of person will say or do probably or necessarily."[12] In order for poetic writing or speech to be probable, according to Aristotle, the poet imitates actions permissible within the bounds of "real" nature, consequently preserving a strong connection between representation and an "actual," if generalized, referent. (The connection itself remains a matter of convention.) This is part of Aristotle's famous limit on poetry; what could be discovered was restricted to the domain of human interaction.

In contrast, sixteenth-century Neoplatonic theories generally held that words were *intrinsically* mimetic, revealing something significant, essential even, about

---

[10]   CSM, vol. 1, p. 4.

[11]   *Ibid.*, vol. 1, p. 39.

[12]   Aristotle, *Poetics*, trans. James Hutton (New York, 1982), p. 54 (1451b).

their referents, regardless of that referent's status as thing or action.[13] Art "reach[es] beyond natural phenomena to the underlying principles of nature," revealing the ideal order behind representations.[14] If language is inherently reflective of what it expresses, then literary production, for the Neoplatonist, is a locus for discovery of the hidden order of the universe. Yet the Neoplatonic text, like the Aristotelian, remains basically mimetic—or "world-imaging," as Czech literary theorist Lubomir Doležel phrases it. Such world-imaging texts give the reader access to the actual world *via* what we now might call the virtual or representational medium of the text.[15]

By contrast, the writer of *possible* worlds remains entirely within the virtual. S/he posits a world that is an *alternative* to the actual world, in its order and origins. As Ruth Ronen suggests, *possible worlds* texts do not make meaning "relative to an extratextual universe."[16] Instead, we see within any possible world "discourse constructing its own world of referents."[17] The imaginary objects to which the propositional discourse of poetic writing refers do not carry a fixed ontological status: they exist neither in physical-material reality, nor in a kind of mental existence. Instead "they remain inherently indeterminate; their state of being is confined to what meaning-units of the text reveal."[18] The indeterminacy of the possible in no way thwarts Descartes's sense of its epistemological value; he is particularly fond of asserting that possible entities are "real," even if they are not "something actually and specifically existing."[19]

The nature of the relationship between the closed system of meaning of possible worlds and other, more open, models of reference in fictional worlds, especially utopias, is a topic of debate amongst modern scholars. According to Mary Baine Campbell, for example, utopian worlds are "not really a version of an other world" because the "normative example" of the utopian world "exert[s] a pressure on readers to alter their own worlds in its direction."[20] By contrast, Marie-Laure Ryan subsumes both possible worlds and "wish-worlds," which include utopias, under the general rubric of "relative worlds"—all of which are relative to some actual world.[21] Campbell's formulation rests on a particular relationship between reader, fictional world, and actual world; put crudely, the text directs the reader to change the world. I suggest reframing our definitions in terms of construction, rather than reception, a move that is more in line with Ryan's work. In texts like More's

---

[13] See James Bono, *The Word of God and the Languages of Man* (Madison, 1995).

[14] Richard Halliwell, *The Aesthetics of Mimesis* (Princeton, 2002), p. 32.

[15] Doležel, *Heterocosmica*, p. 2.

[16] Ronen, *Possible Worlds*, p. 41.

[17] *Ibid.*, p. 29.

[18] *Ibid.*, pp. 98–9.

[19] *Ibid.*

[20] Campbell, *Wonder and Science*, p. 15.

[21] Marie-Laure Ryan, "The Modal Structure of Narrative Universes," *Poetics Today*, 6 (1985): pp. 717–55.

Utopia (1516), or Bacon's *New Atlantis* (1627), the authors pull from society "the rules it has prescribed itself in different circumstances, select what appears to them the wisest, and linking them together in an imaginary vinculum, give birth to a state, a form of government, a code of laws, and a system of manners, such as in their totality never existed"—an Aristotelian form of mimesis.[22] In contrast, *The World*, according to Descartes, is not abstracted from the actual in order to produce the ideal; rather, it produces an alternative possible world from which lessons in mechanics, metaphysics, and optics can be learned.

While possible worlds are created as alternatives to the actual, this is not to say that possible worlds are not accessible (in a logical sense) to the actual. Indeed, the epistemological power of *The World* depends on such accessibility. Ryan argues that such self-enclosed narratives retain a radical otherness to the actual while maintaining an "epistemic or model function."[23] Descartes spatially produces Ryan's "radical otherness" for *The World*, while simultaneously calling attention to its crafted or closed nature. The reader is exhorted to "allow your thought to wander beyond this world to view another, wholly new, world."[24] This wandering takes the reader "far enough to lose sight" of the actual world, suggesting a space not only other, but so distant as to prohibit proximate comparison. The crafted nature of this other world is signaled by Descartes's rhetoric of invention and creation; the new world is "called forth" by Descartes into an imaginary space "invented" by philosophers.[25] Together, reader and author spend pages inventing, imagining, and conceiving of a different (alternate, in Ryan's terms) possible world. While this world is "a most perfect world," Descartes continues to keep it discrete in terms of both creation and reference, leaving it to others to "explain the things that are in fact in the actual world."[26] While in the case of *The World* a comparative return to the actual is the final goal—Descartes wants to demonstrate that an alternative, mechanically created world could look just like our own—the comparison is always between two distinct worlds.

Discovery in *The World*

Words can write worlds into being. Yet in *The World* it is God's ability, in principle, to bring a given "paper world" into material being that gives such a world—a text—its epistemological value. Discovery, in this scenario, is not of what is, but of what may be: a possibility of God's creative capacity to match the creative writing of man. Descartes is clear that his possible world does not challenge received

---

[22]   J.A. St. John, *Utopia: or the happy republic. To which is added, The New Atlantis, by Lord Bacon, ... by J.A. St. John, Esq.* (London, 1845), p. iv.

[23]   Ryan, "Modal Structure," p. 730.

[24]   Descartes, *The World*, p. 21.

[25]   *Ibid.*, p. 21.

[26]   *Ibid.*, p. 24.

histories of creation. Neither *The World*, nor its "real" but non-existent referent, can be collapsed onto the actual world. The "fable" does "not ... explain the things that are in fact in the actual world."[27] Instead, it allows Descartes to "make up ... a world in which there is nothing that the dullest minds cannot conceive;" a world as clear and distinct as intuition.[28] While creation may not have happened in "the old world" (i.e. the actual world) as it does in Descartes's possible world, nevertheless, according to Descartes, this world *is possible*: "since everything ... here can be imagined distinctly" and "it is certain that God can create everything we imagine."[29]

According to *modern* theories of possible worlds, writing the possible entails creation in thought and language: stipulation rather than discovery. The veracity of such writing is subject only to systems or regulations imposed by the writer and reader.[30] In *The World*, the regulations are Cartesian intuitions. While these intuitions constitute a kind of discovery in Cartesian philosophy ("intuitions" are for Descartes truths subject to discovery) *The World* is a text that *stipulates* in order to discover. From intuition Descartes "make[s] up" a world.[31] The active, creative force of "makes up" signals stipulation, which is then further refined and reinforced when Descartes suggests that even if God has not created a world as it is represented here, it is enough that he *could*. As a devout Catholic, Descartes was not in a position to argue that he was the actual author of a possible world. Instead, he wrote an account of what is possible for God to make. The reader and Descartes "suppose that *God* creates anew so much matter all around us," not that *they* do so.[32] While the reader and author together "imagine" and "conceive of" this world "as [they] fancy," it is nevertheless God who establishes its laws, divides its matter into parts, and sets the universe in motion.[33] While Descartes has "imagined" and made up this world "as [he] pleased," its certainty, and therefore its intellectual utility, is based in God's creative power; even if he has not made this world already, "he can create everything we imagine."[34] As Descartes says, what is "true or possible" is such only because "God knows [it] to be true or possible."[35] Consequently, rather than operating according to the modern sense of

---

27   *Ibid.*

28   *Ibid.*

29   *Ibid.*

30   Citing Kripke, Ronen notes: "possible worlds are 'stipulated not discovered by powerful telescopes'." She argues against thinking of possible worlds as newly discovered mental lands, as that would attribute a kind of "thereness" to them that would be contradictory. See Ronen, *Possible Worlds*, p. 103.

31   CSM, vol. 1, p. 256.

32   Descartes, *The World*, pp. 21–2.

33   *Ibid.*, pp. 22–3.

34   *Ibid.*, p. 24.

35   CSM, vol. 3, p. 24.

creative "stipulation," *The World* is a stipulation of divine creation, and therefore an object of discovery.

This imaginative stipulation is cast in the same terms used throughout Descartes's corpus to describe his method of "discovery." It is just as certain as empirical or observational discovery: everything in this divinely-created new universe can be known through the narrative "so perfectly" that his readers cannot even "pretend to be ignorant of it."[36] What the reader cannot be "ignorant of" is the possibility of a different divine creation than the one attested to by and for the Earth. The narrative offers knowledge of divine capacity, not necessarily material reality. Interestingly enough, it is only through a careful and precise process of *imagination* that this discovery of divine creative capacity is possible. Imagination is necessary because the possible world is not written from within the context of the material world.

The Possible World as Intuited

Possible worlds are unified and coherent, and this feature was important to Descartes's own sense of *The World*'s epistemological value. His theory of intuitions often requires that the investigator reduce the question at hand to the most basic level in order to achieve ideas that are clear and distinct. Nonetheless, Descartes also allows that more complicated ideas may be *compound intuitions*, which are just as valid as the simples from which they are constructed.

A 1648 exchange recorded by Dutch scholar Frans Burman illuminates how compound intuitions authorize possible entities. Frustrated at Descartes's insistence that "imaginary" mathematical entities like polygons were "real," even if not present in natural creation, Burman challenged Descartes with the example of the chimera. For Burman, a *poetic* figure like the chimera—described in Homer's *Iliad* as a horrible hybrid animal, "lion-fronted and snake behind [and] a goat in the middle"—could be neither real, nor epistemologically useful.[37] On the contrary, Descartes retorted: "everything in a chimera that can be clearly and distinctly perceived is a true entity."[38] Consequently, it was *possible* for the compound of those clear and distinct parts likewise to have a "real and true nature."[39]

For Descartes, clear perception enabled by the motion of the imagination could qualify the chimera as "something which does not actually exist in space but is capable of so doing."[40] This was different from "clearly imagining" the chimera: "even though we can with the utmost clarity imagine the head of a lion joined

---

[36]   Descartes, *The World*, p. 23.

[37]   Homer, *The Iliad of Homer*, trans. Richard Lattimore (Chicago, 1961), bk 6, ll. 180–82.

[38]   CSM, vol. 3, p. 343.

[39]   *Ibid.*

[40]   *Ibid.*

to the body of a goat," Descartes tells Burman, "it does not therefore follow that they exist, *since we do not clearly perceive the link*, so to speak, which joins the parts together."[41] Imagination *aids* the intellect toward clear perception of not just the featured elements, but also the nature of their juncture. It was not enough to imagine that a lion's head and goat's body could go together; one had to be able to understand *how* they went together as well.

In emphasizing linkages we can see Descartes's prioritization of a "simultaneous cognition" of knowledge. He is looking for the unifying principle that constitutes poetic knowledge rather than individual truths in succession.[42] When dealing with multiple connected ideas, as in the case of the chimera or of a possible world, there must be what Descartes describes in the *Rules* as a "continuous movement of thought," enabled by the imagination, in order to unify the separate parts of a more complicated intuition.[43] The regulating order of the Cartesian possible world must remain unified, coherent, and everywhere apparent; the linkages of the Cartesian imagination "must nowhere be interrupted," lest the entire narrative be "immediately broken" and its certainty collapse.[44] Not that the imagination's aid is always salutary; the imagination may put "things together poorly" and, in so doing, "makes us imagine many events are possible when they are not."[45] Thus the chimera would be a fiction if the links between goat and lion were not present. But when the links are made clearly, such that interconnections are as clear and distinct as the ideas, then poetic knowledge is a kind of all-at-once cognition of both the nature of the unifying principle(s) and an appreciation for the beauty of the system.[46] While fictions fail (epistemologically) because of poor construction, coherent possible worlds successfully enable discovery.

The World with its mechanistic "links" may be only possible, and, therefore, imaginary, but for Descartes, it is not "fictitious."[47] Indeed, Descartes's concern to preserve the conceptual unity and validity of *The World* prompted him to suppress the text, despite his obvious affection for its argument. In April 1634 Descartes wrote to the Minim mathematician Martin Mersenne, apologizing for his delay in sending a long-promised copy of *The World*. Citing the prosecution of Galileo for heresy for his 1632 *Dialogue Concerning the Two Chief World Systems*, which included a heliocentric hypothesis, Descartes informed Mersenne that he refused to release his text. The crucial point for Descartes was the need to preserve the unity of the narrative of *The World*: "all of the things I explained in my treatise,

---

41   *Ibid.*, vol. 3, pp. 343–4; emphasis mine.

42   Jones, *The Good Life*, p. 58.

43   CSM, vol. 1, p. 25; on the same page: a "continuous movement of the imagination, simultaneously intuiting a relation and passing on to the next," enables Descartes "to intuit the whole thing at once."

44   *Ibid.*

45   *Ibid.*, vol. 1, p. 114.

46   Jones, *The Good Life*, p. 63.

47   CSM, vol. 3, p. 343.

which include the doctrine of the movement of the earth, were so interdependent that it is enough to discover one of them false to know that all the arguments I was using are unsound."[48] With the prosecution of Galileo, it seemed certain to Descartes that the heliocentrism of his possible universe would likewise come under attack and that such an attack would undermine the entire text. The issue was not heliocentrism, *per se*, but Descartes's belief that a single false link would render the epistemological value of his world null. If heliocentrism were refuted, Descartes's possible world would cease to be coherent and unified, and the text would become a *mere* fiction.[49] He had designed his mechanistic "fable" to produce the simultaneous imaginative motions that would enable his readers *to intuit*, and therefore *to know*, the possibility of mechanistic creation.[50] Afraid that a challenge from the Church would disrupt his delicate, interconnected narrative, Descartes chose not to publish the text.[51]

Cartesian Possibility and Aesthetics

Possible worlds, suggests Thomas Martin, are particularly satisfying objects of knowledge, because they are "eminently more knowable than our own enigmatic lives."[52] They enable the complete knowledge that eludes encapsulation in the actual world. To Descartes, *The World*'s poetic knowledge was so clear that it would allow "even the dullest minds" to grasp the possibility of mechanist creation.[53] This knowledge functioned at two levels. First, the textual model demonstrated its own possibility. Prior to reading *The World*, the reader might have "taken [Cartesian light mechanisms] to be very paradoxical," but after reading they become "clear" and visible.[54] Similarly, readers initially may have had "doubts" about the mechanisms of mortal bodies; but after reading, "they will not be able to imagine anything more likely" than Descartes's model.[55] Second, Descartes designed the text to teach the reader to pick out analogous mechanistic principles at work in the real world. By offering a simplified model, the text

---

48   *Ibid.*, vol. 3, p. 42.

49   Descartes, *The World*, p. 6.

50   *Ibid.*, p. 21.

51   Readings that argue that Descartes's use of possibility was either a political and disingenuous hedge or is simply not germane, such as Stephen Fallon's argument that the fiction of the possible world and scenarios is "itself a fiction," a "strategic evasion," misunderstand the utility Descartes sees in poetic knowledge, as well as the strict limitations he places upon it. Such readings elide the importance of imaginative narratives to Cartesian philosophy. See Stephen Fallon, *Milton among the Philosophers* (Ithaca, 1991), p. 29.

52   Martin, *Poeisis*, p. 141.

53   Descartes, *The World*, p. 24.

54   *Ibid.*, p. 70.

55   *Ibid.*, p. 168.

teaches the reader to cut through the noise of the complicated real world, in order to discover mechanist principles at work. Discovery is a two-stage process, then, that begins in the text and continues in the world. The clarity of that perspective in the *possible* world is what lends authority to a translated perspective in the *actual* world. In some ways this is utterly conventional. Learning how to read the natural world through a text was extremely common. What is different here is the level of certainty attributed to textual evidence and the elision of the actual world as the authorizing reference for the text. There is no recourse to empirical proof here—because Descartes's mechanistic world is only *possible*, the text alone must be evidence enough. The turn to the actual world does not verify the text; rather, it deploys the knowledge gained therein in a comparative exercise. While not an outright rejection of empirical proof, Descartes's world is an alternative approach to discovery that places imaginative narrative—the aesthetics of which was itself evidence of the epistemological value of the story—at its center.

It is worth noting that this text is notable not only for its location in the history of poesis and possible worlds, but also because it provokes a rethinking of certain generic distinctions within possible-worlds theory itself. For Doležel, the possible worlds of fiction are aesthetic artifacts. Those of natural science are practical, non-aestheticized models of alternative worlds. Descartes, however, values his possible world in both aesthetic and pragmatic terms. As we will see, his possible world is simultaneously fiction and natural science (in Doležel's formulation). As such, it is yet another reminder to us that "discovery," at least in the early-modern period, can be as much about aesthetics as it is about knowledge. In troubling the distinction between "stipulation" and "discovery," which is at the heart of modern possible-worlds theory, *The World* also illuminates a historical moment in which the line between creative agency in the author, and discovery of a divinely created possibility, is blurred. Descartes cannot claim that the written word actually creates new material worlds. He can and does claim, however, that texts can articulate what has not yet been created; that, in addition to their ability to describe the world, they can explore the contours of divine possibility.

In building a radically other world, Descartes simplifies the natural order of his universe by rejecting the paradigm of Aristotelian qualities and the traditional four elements and substituting for them a tri-elemental universe devoid of all characteristics, except those of motion and extension. While complex ideas, such as possible worlds, can qualify as intuitions if the linkages are clear, Descartes strongly privileges focusing on "minute details" first.[56] Accordingly, *The World*'s first five chapters focus on topics of a far smaller scale than world-creation, thereby serving both a methodological and narratological function. In the *Rules*, Descartes argues that

---

[56] Rule nine of the *Rules* suggests that inquiry must begin with "the most insignificant and easiest of matters." CSM, vol. 1, p. 33.

we should not begin ... by investigating difficult matters. Before tackling any specific problems we ought first to make a random selection of truths which happen to be at hand, and ought then to see whether we can deduce some other truths from them step by step, and from these still others, and so on in logical sequence. This done, we should reflect attentively on the truths we have discovered and carefully consider why it was we were able to discover some of these truths sooner and more easily than others, and what these truths are. This will enable us to judge, when tackling specific problems, what points we may usefully concentrate on discovering first.[57]

What Timothy Reiss has called Descartes's "aesthetic rationalism" is apparent here.[58] Inquiry begins with a deductive sequence and ends in a synthesis that enables, through the ability to "judge," further discovery. By reflecting on the knowledge obtained from a purportedly random selection of "truths," Descartes and his reader can recognize the interconnections that enable the discovery of order ("why it was we were able ..."). Recognition of interconnection enables judgment, making further discovery an aesthetic (based in judgment) rational process. What Descartes describes in the *Rules* as "sound judgment" enables those of sufficient intellectual capacity to make new discoveries. And for the general public, it also makes possible an aesthetic assessment of the sciences as a whole: "even those who, through lack of intelligence, cannot make discoveries by employing first principles will still be able to recognize the true worth of the sciences, and this will enable them to arrive at a correct judgment of the value of things."[59]

Descartes opens *The World* with possibility: "I want to draw to your attention that it is possible for there to be a difference" between the sensation of light and the cause of that sensation. Built into Descartes's suggestion is his familiar skepticism about the senses. But here it is cast in terms of an alternate possible state of affairs: when looking at light what we think we see is not the cause of our perception of light. Descartes offers two analogies to illuminate the interpretive nature of the error, the first of which will suffice here. As with conventional language, Descartes argues, it is possible that the idea of light is no more connected to the sign of light than the mechanics of speaking are connected to the content conveyed by speech. Descartes asks:

> now if words, which signify something only through human convention, are sufficient to make us think of things to which they bear no resemblance, why could not Nature also have established some sign which would make us have a sensation of light, even if the sign had in it nothing that resembled this?[60]

---

[57]   *Ibid.*, vol. 1, p. 23.

[58]   Timothy J. Reiss, *Knowledge, Discovery and Imagination in Early-Modern Europe* (Cambridge, 1997).

[59]   CSM, vol. 1, p. 3.

[60]   *Ibid.*, vol. 1, p. 4.

Descartes suggests that light is nature's conventional sign for motion. The error is in the relationship of sign to signified; Descartes argues that this relationship is conventional rather than necessary or one of identity. Interpretation becomes the central issue, and the possibility of possibility is the occasion to reconsider the nature of interpretation. Descartes argues that it is possible that the natural world is a system of signs subject both to convention (hence, arbitrary signification) and to misinterpretation. In Descartes's possible world, the book of nature has to be read with as much interpretive effort as any other conventional-language text. At this early stage in the text, Descartes is only proposing possible scenarios in the real world, not entire new worlds. Nevertheless, this introduction owes as much to the logic of possibility, and its epistemic value, as it owes to Cartesian theories of light. By reforming light into a conventional sign, Descartes asks his readers not to be convinced "absolutely that light is something different in objects from what it is in our eyes"; he wants "only to raise a doubt about it for [us], to prevent [our] being biased in favor of the contrary view, so that we can examine what light is."[61] It is under the condition of possibility that Descartes posits further collaborative inquiry.[62]

The reader is thus introduced to *possibility* as the enabling condition of knowledge. It is only once we, as readers, accept the possibility that light is not what we think it is that real discovery begins. This is fundamentally different from reading this possibility within a tradition of probabilistic rhetoric that allowed early-modern writers to skirt charges of heresy or to massage the non-syllogistic rigor of their propositions. While the possible scenarios proposed in the first five chapters may be usefully read in the context of probability, the larger "fable" that Descartes introduces at the end of Chapter Five does not fit easily into this paradigm, nor does the context of early-modern probability adequately account for the non-mimetic mode Descartes uses to articulate his possible world.

The selection of topics that begin *The World*, including light and the nature of motion, cannot reasonably be considered "random"; Descartes rarely begins with anything that could be considered truly random. Instead, he often begins with what appear to be arbitrary selections that, in the end, serve the goal of each proof precisely, as he does with the first five chapters of *The World*. Chapters One and Four introduce possibility as a condition for inquiry, and the five "introductory" chapters model the methodology laid out by Descartes in *The Rules*. At the same time, they establish the fundamental elements at the heart of the possible world, thereby establishing the certainty and aesthetic value of the alternative principles essential to the coherence of Descartes's narrative.

Descartes eliminates all "superfluous" data in order to arrive at the simple principles upon which his alternative world will be built.[63] Chapter One establishes

---

61   *Ibid.*, vol. 1, p. 6.

62   See Sylvie Romanowski, "Descartes: From Science to Discourse," *Yale French Studies*, 49 (1974): pp. 96–109.

63   CSM, vol. 1, p. 28.

possibility as a condition for inquiry and disabuses readers of the bias that the sensation of light is connected necessarily to the cause of that sensation. Chapter Two continues the discussion of light by defining the nature of light in fire or stars in terms of motion. Chapter Three presents a theory of the conservation of total universal motion in the course of a discussion of "hardness" and "fluidity." This, in turn, places all bodies along a fluid–solid state spectrum defined in terms of motion. We begin to see, at this point, the nature of the interconnection of the "random" selection of light, fire, "hardness," and "fluidity": Descartes understands them all in terms of motion. Chapter Four seems to shift focus somewhat, opening with a question of perception related to that in the first chapter and reintroducing the problem of bias that was at issue in that chapter as well. "But we need to examine in greater detail," Descartes tells us, "why, although it is as much a body as any other, air cannot be sensed as easily."[64] In the ensuing discussion, Descartes argues that the universe is a plenum (vacuums do not exist), and planetary and other celestial bodies float in a rarified fluid that we cannot see because we are surrounded by it. Though initially about perception, this chapter argues that motion must be circulatory (all bodies must be in motion in the same direction in order to open the space into which any given body will move if there is no vacuum or empty space). Descartes, at this point, has defined light in terms of motion, determined that all bodies exist along a solid–fluid spectrum defined in terms of motion, and established the kind of motion possible given that the universe is a plenum—the "links" that will make this story intellectually and imaginatively clear are "made" of motion. What remains is to define the elemental components that will move in this universe.

This is the project of Chapter Five, in which Descartes reduces the traditional four elements (fire, water, earth, air) to three (fire, earth, air). He defines each in terms of motion and extension (shape and size), thus continuing the mechanical emphasis of Chapters One through Four. This move also disposes of the Aristotelian qualities of hot, cold, wet, and dry that traditionally accompanied the four elemental natures and, in Descartes's opinion, cluttered philosophical inquiry. Descartes then arranges the three elements in a mechanical hierarchy. "Fire" is composed of tiny entities with great speed and "no determinate size, shape or position" (they fill in the space between larger bodies, ensuring that the universe is a plenum).[65] "Air" is of "middling" motion and size, with a spherical shape, and "earth" is "so large and closely joined ... that [its particles] can resist the force of motion in other bodies."[66] Descartes distinguishes elemental air, fire, and earth from the "gross air [that] we breathe or the earth on which we walk."[67] With this last chapter Descartes

---

64 Descartes, *The World*, p. 13.

65 *Ibid.*, pp. 17–18.

66 *Ibid.*, p. 18.

67 *Ibid.*, p. 19.

closes the introductory material of *The World*, thus completing the "ordering and arranging of the objects on which we must concentrate."[68]

In reducing the complicated problem of the construction of the universe to the motion and extension of three clearly defined elements, Descartes has employed possibility as a device to overcome anticipated biases (that what we perceive as light is the same in the bodies in which we perceive it, and that where we see nothing, nothing is present). But it is not enough simply to have found motion in the actual world; Descartes's objective is to demonstrate that God could have used these principles of motion to create out of "chaos."[69] And to demonstrate that even "if everything our senses ever experienced in the actual world seemed manifestly contrary" to this mechanistic method of creation, the certainty of these principles "in the new world" is so "infallible" and "perfect" that the reader must acknowledge that such creation is *possible*.[70] This introductory material, then, models the Cartesian method and articulates the principles of inclusion and exclusion that govern the narrative. This clear, distinct, closed, and completely coherent narrative enables a new poetic world to emerge.

The "End" of the Possible World: Analogy

Writing about this new possible world served two functions for Descartes. First, it allowed him to put forth a model of a world constructed "without forms or qualities," one based solely on a modified fluid mechanics.[71] It is an "alternative design of the universe constructed by varying the basic physical constants"—constants such as basic elements, the nature of light, and so on.[72] While the cohesive narrative qualified the possible world as certain as intuition, another part of its value lay in the ability to argue by analogy for similar mechanisms in the actual world. Descartes's possible new world is one "in which one will be able to see not only light, but all other things as well ... that appear in the actual world."[73] Understanding the possible world allowed the reader to perceive *correctly* phenomena in both worlds. By eliminating the multitude of other considerations and stripping down to a model of creation based solely on motion, Descartes sought to enable his readers to conceive "distinctly" the mechanical principles behind material phenomena. Mechanistic principles may "be as true in each [of God's worlds] as in this [possible] one."[74] What happens to Descartes's imaginary

---

[68]   CSM, vol. 1, p. 20.

[69]   Descartes, *The World*, p. 25.

[70]   *Ibid.*, pp. 28, 31, 23.

[71]   *Ibid.*, p. 22.

[72]   Doležel, *Heterocosmica*, p. 14.

[73]   Descartes, *The World*, p. 23.

[74]   *Ibid.*, p. 31.

men and animals in this new universe *can* happen "in the actual world as well."[75] Consequently, his "paper world" functioned as a locus for the development of clear and distinct understanding of mechanist principles that enabled the reader to judge for him/herself if, while created differently, these mechanisms might have analogs in the actual world.

The second function that this possible world served was *aesthetic*. According to Descartes, he used the "fable" of the possible world "so as to make this long discourse less boring." He wrote a "pleasing" story in order to enable "truth" to "manifest itself sufficiently clearly."[76] Recall that, for Descartes, the value of poetic knowledge lay in the ability of the imagination to grasp the individual components of a narrative in an all-at-once kind of intuition (here of motion), thus enabling clear and distinct knowledge of a more complicated sort. As we saw earlier, the well-executed narrative of discovery teaches strong intellects the judgment needed for new discovery, and weaker intellects an aesthetic appreciation for such mental work. "Wrapping up" the "naked" philosophy made it possible for Descartes to present his reader with a unified and brightly shining alternative world.[77] Descartes had hoped that this alternative world would be "useful to some," in so far as it presents an illuminating possibility and teaches the judgment necessary to undertake further discovery.[78] For others, it was to have demonstrated the clarity and rigor of such imaginative understanding. While analogical discovery was clearly important to Descartes and the aesthetics of poetic knowledge gesture directly toward "weaker" intellects, there is a sense of excitement in the text that suggests that Descartes saw real innovation in poetically generated knowledge. *The World* "pleases" through "shading and bright colors," while making the nature of the epistemological "links" clearer than is possible in mimetic discourse. Descartes believed his fable facilitated higher-order knowledge and aesthetic judgment.[79] His recognition of the danger posed by his "paper-world" speaks not only to the intellectual milieu of the mid-seventeenth century but also to the profound power he saw in poetic knowledge. Though he recognized the danger of such propositional writing, Descartes was "too much in love" with the beauty and rigor of *The World* completely to suppress the text. His new universe periodically appears in his more conventional and mimetic texts, a constant reminder throughout his empirical work that the realms of the possible were waiting to be discovered.

---

[75]   *Ibid.*, p. 70.

[76]   *Ibid.*, p. 21.

[77]   *Ibid.*, p. 21.

[78]   CSM, vol. 1, p. 112.

[79]   Descartes, *The World*, p. 32.

Chapter 8

Numbering Martyrs: Numerology, Encyclopedism, and the Invention of Immanent Events in John Foxe's *Actes and Monuments*

Ryan Netzley

Non poena sed causa: the cause, not the penalty, makes the martyr. All early-modern martyrologies, Protestant and Roman Catholic, at least pay lip service to this notion. Readers and even witnesses must discover—and ultimately approve—the cause for which someone dies before enrolling him or her in the ranks of the faithful. The problem with this imperative, of course, resides in the fundamentally inaccessible nature of interior motivations; which, as Milton famously notes, run the risk of authorizing any obstinate heretic as a martyr. "If to die for *the testimony of his own conscience* be anough to make him Martyr," Milton writes, "what Heretic dying for direct blasphemie, as som have don constantly, may not boast a Martyrdom?"[1] In other words, it is not just that the deaths of martyrs and heretics look similar. Their internal commitments and consciences are also analogous. To tell the difference between them, one would need a notion of cause that is not secreted in the psychic recesses of individuals.

Luckily, early-modern martyrology includes just such a notion. "Cause," in this register, denotes an immediately apprehensible—and, as it were, already discovered—testament to the truth of one's faith. After all, in classical rhetoric, *martyria* (from the Greek *marturion*, or testimony) is a figure of speech in which a witness "confirms something by virtue of his/her own experience."[2] In other words, to be a true martyr, one must possess a discoverable cause, true faith, but must also manifest this faith; so that its unearthing becomes, at least on one level, unnecessary. Martyrologies like John Foxe's *Actes and Monuments* (1583) hold out the promise of and demand for discovery—of discerning precisely the true faith behind the suffering and death of individual martyrs. Yet they can also insist that the martyrological discovery has already occurred, because it has

---

[1]   John Milton, *Eikonoklastes*, in *Complete Prose Works of John Milton*, ed. Merritt Y. Hughes (New Haven, 1962), vol. 3, p. 576.

[2]   Susannah Brietz Monta, *Martyrdom and Literature in Early Modern England* (Cambridge, 2005), p. 4.

been made manifest in the world precisely as and via testimony. Thus Foxe's encyclopedic catalogue of Protestant and proto-Protestant martyrs (not just of the Marian persecutions, but throughout church history) is not an investigation into the truthfulness of the martyrs's faiths: that determination occurs before, or as a precondition of, any given martyr's inclusion in the volume. Foxe is not revealing the valid causes behind his martyrs's deaths. Rather, he is listing the valid deaths that follow from such causes.

The resulting problem, obviously, is a potential violation of the principle with which martyrology begins: *non poena sed causa*. If we are not reading martyrological narratives for ethical or intentional proof that a given martyr truly was one, what are we reading them for? A mere enumeration or listing of martyrs, catalogued for their similar and grisly deaths, runs the risk of producing a readership sitting in awe or terror of the spectacular death itself, instead of attempting to discern aright the cause for which a given martyr died. Nicholas Harpsfield, Foxe's chief Roman Catholic antagonist, uses *non poena sed causa* as a means of criticizing Foxe's compendious, paratactic inclusiveness: "We oppose Foxe, not with the number of martyrs, but rather with their weight, not with their deaths, but with the causes of there deaths."[3] Foxe, of course, would insist that all of the martyrs included in the *Actes and Monuments* carry such weight. Yet Harpsfield's condemnation is interesting insofar as it implies that the size of the book might actually detract from an attention to the cause for which the martyrs died. That his critique of Foxe remains couched in metaphors of measurement—weight instead of enumeration— only highlights the pivotal role that number seems to play in martyrologies and martyrological debates.

This chapter seeks to explore Foxe's uneasy combination of discovery and transparent manifestation: of martyrdom conceived as the discovery of the truth of a given martyr's intentional cause, and *martyria* conceived as confirmation via transparent witness. It also attempts to take seriously Harpsfield's critique, and to consider, more broadly, the ways in which number, chronology, and the encyclopedic ambitions of the *Actes and Monuments* interact with notions of invention and discovery. Foxe's impulse to comprehensiveness, and the propensity for repetition, if not repetitiveness, that goes along with it, may actually be how this massive tome functions as a text. As Mark Breitenberg notes:

> The *Acts and Monuments* cannot have been read or heard as a linear narrative progressively imparting new information or events; the outcome of these stories was probably already known through other sources, and certainly known to

---

[3] Nicholas Harpsfield, *Dialogi sex contra summi pontificatus, monasticae vitae, sanctorum sacrarum imaginum oppugnatores, et pseudomartyres* (Antwerp, 1566), p. 992. Quoted and translated in Thomas S. Freeman, "Introduction: Over Their Dead Bodies: Concepts of Martyrdom in Late Medieval and Early Modern England," in Thomas S. Freeman and Thomas F. Mayer (eds), *Martyrs and Martyrdom in England c. 1400–1700*, *Studies in Modern British Religious History*, 15 (Woodbridge, 2007), p. 21.

anyone who had read at least one of Foxe's accounts before. Thus, it is not so much what happens or what is said, but how often it is repeated ... repetition enables Foxe to heap his stories and documents on top of each other in an effort to amass an entire Protestant "state" of texts in his book, from royal proclamations to village conversations.[4]

Foxe does not offer us revelation or discovery at the end of each martyr's narrative, or the book as a whole. Neither does the *Actes and Monuments* simply accumulate evidence—all of those documents, letters, proclamations, and narratives—for a partisan debate about the legitimacy of the Protestant church. Instead, the aim of Foxe's compendium, and of the numerical, numerological, and chronological machinations that go along with it, is precisely to avoid debates about the relationship between determinism and freedom, discovery and invention. The force of the form—the comprehensive Protestant encyclopedia—transforms what we can mean by a cause or commitment: it is no longer a reason or motivation hidden in the secret recesses of a martyr's soul, but an explicitly manifest, textual event. The point I wish to emphasize is that repetition and the impulse to comprehensiveness, in Foxe's martyrology, produces an interest in number that fundamentally changes how we imagine the relationship between events, even traumatic events, and books. This relationship, it turns out, revolves around a host of invented numbers, and the manner in which they mean.

Inventing Numbers

On the very first page of the text proper of the *Actes and Monuments*, Foxe lays out a chronology of five ecclesiastical epochs: (1) 300 years of suffering, including the time of the apostles; (2) 300 years of flourishing, amongst the early church fathers; (3) another 300 years of backsliding; (4) 400 years of the time of Antichrist; (5) 280 years of reformation. It is this last era, and Foxe's reticence about determining its endpoint, that matters most for us initially: "The durance of which tyme [the time of reformation] hath continued hetherto about the space of 280. yeres, and how long shall continue more, the Lord and gouernour of all tymes, he onely knoweth."[5] Despite this avowed reticence to appoint the time of the apocalypse, Foxe has already done so in one of the book's prefatory sections, "Foure Questions

4   Mark Breitenberg, "The Flesh Made Word: Foxe's *Acts and Monuments*," *Renaissance and Reformation*, 15 (1989): pp. 381–407; p. 392. See also D.R. Woolf, "The Rhetoric of Martyrdom: Generic Contradiction and Narrative Strategy in John Foxe's *Acts and Monuments*," in Thomas F. Mayer and D.R. Woolf (eds), *The Rhetorics of Life-Writing in Early Modern Europe: From Cassandra Fedele to Louis XIV* (Ann Arbor, 1995), pp. 243–82; p. 258.

5   John Foxe, *Acts and Monuments [...] The Variorum Edition* (http://www.hrionline. shef.ac.uk/foxe/), 1583 ed., bk 1, p. 1. Unless otherwise noted, all references are to the

Propounded to the Papists." Here, he hints at a much more specific dating of the end of this fifth age—294 years—and attempts to identify the two beasts of Revelation with the papacy by highlighting a historical parallel between the duration of the early and recent persecutions:

> Thirdly, where the sayd beast had power to make 42. monethes and to fight against the Saintes, and to ouercome them, &c. therby most manifestly is declared the Empyre of Rome, with the heathen persecuting Emperours, whiche had power geuē the space of so many monthes, (that is, from Titherius to Licinius. 294. yeares) to persecute Christs Church as in the Table of the primitiue hereafter following is discoursed more at large.[6]

As Andrew Penny notes, Foxe never can "avoid the temptation to get involved with dates"; Katharine Firth describes Foxe as unique amongst British commentators for his attempt to reconcile the numbers in the Book of Revelation with the events and durations of history, instead of just using John's apocalypse to identify Rome and Antichrist.[7] This early section of the *Actes and Monuments* is certainly an example of these propensities. The preface equates 42 months and 294 years so as to bring the numbers from the Book of Revelation into conformity with historical events in the late-medieval and early-modern period, and calculate the length of the remaining period of the true church's persecution. Yet Foxe also defers explanation of this equation for over 100 pages. Both in the text and in the margin, Foxe alerts readers that the link between 42 and 294 will be "discoursed more at large"; the marginal note points readers to the page on which this later exposition occurs.[8] What is important about this moment is that the numbers that one might expect to be an explanatory device become an excuse for discovery and mystery in Foxe's text—which is otherwise committed to the repetition of similar, even tiresome, narratives that always lead to the same point: the martyrs's deaths.

Not only does 294 appear early in the *Actes and Monuments* as a mysterious number in need of further and later explanation; when Foxe actually explains it, the number appears as transparently invented, and in a dream at that. Foxe turns to the Book of Revelation in an attempt to determine the length of the present period of persecution, and provide solace to the godly victims thereof: "Onely in these

---

1583 edition. For Foxe's prefaces, this chapter uses the revised pagination of the *Variorum Edition*.

[6]   Foxe, *Actes*, Pref., p. 18.

[7]   Andrew Penny, "John Foxe, the *Acts and Monuments* and the Development of Prophetic Interpretation," in David Loades (ed.), *John Foxe and the English Reformation* (Aldershot and Brookfield, 1997), pp. 252–77; p. 264; and Katharine R. Firth, *The Apocalyptic Tradition in Reformation Britain, 1530–1645* (Oxford, 1979), p. 91. See also Palle J. Olsen, "Was John Foxe a Millenarian?" *Journal of Ecclesiastical History*, 45 (October 1994): pp. 600–624.

[8]   Foxe, *Actes*, Pref., p. 18.

persecutions I could finde no end determined," he writes, "nor limitation set for their deliverance."[9] The problem is that the almost 300 years of recent persecution for which Foxe's history must account do not correspond to any of the pivotal apocalyptic durations in John's apocalypse: 42 months, 1260 days, three and a half years, and three and a half days. Foxe's solution to this conundrum comes to him in a dream:

> Thus being vexed and turmoyled in spirite, about the reckening of these numbers and yeares, it so happened upon a Sonday in the morning lying in my bed, and musing about these numbers, sodenly it answered to my minde, as with a majestie, thus inwardly saying within me: thou foole count these moneths by Sabbots, as the weekes of Daniell are counted by Sabbots. The Lord I take to witnes, thus it was. Wherupon thus being admonished, I began to recken the 42 monethes by Sabbats, first of monethes, that would not serve, then by Sabbots of yeres wherin I began to feele some probable understanding.[10]

Foxe invents the number 294 by multiplying 42 months—the time period allotted to the beast to speak blasphemies (Revelation 11:2 and 13:5)—by seven, what he dubs "counting by Sabbots." $42 \times 7 = 294$. Yet the invention does not stop there: he must also change "months" to "years" to fit the historical periodization that he is attempting. Strangely, Foxe then reports verifying the basic calculation of 42 times seven with several merchant friends: "Yet not satisfied herewith, to have the matter more sure, eftsoones I repaired to certain Merchants, of myne acquaintance … To whom the number of these foresaid 42 monethes, being propounded and examined by Sabbots of yeares, the whole summe was found to surmount to 294 yeres, conteining the full and just tyme of these foresaid persecutions neither more nor lesse."[11]

At issue here may be the numerological significance of 294: *Elohi Abraham*, The God of Abraham.[12] The *Actes and Monuments* never mentions, much less endorses such a meaning, yet Foxe exhibits substantial familiarity with numerology in *Eicasmi seu meditationes in sacram Apocalypsin*, his still untranslated commentary on the Book of Revelation. This text includes a table identifying the Pope and the Antichrist via the numerological values of both Hebrew and Greek

---

[9] Foxe, *Actes*, bk 1, p. 100.

[10] *Ibid.*

[11] *Ibid.*, pp. 100–101.

[12] David Godwin, *Godwin's Cabalistic Encyclopedia: A Complete Guide to Cabalistic Magick*, 3rd ed. (St. Paul, 2003), p. 542. For brief overviews of the numerical values of Hebrew letters and gematria charts, see *Godwin's Cabalistic Encyclopedia*, pp. 645–8; Karl Menninger, *Number Words and Number Symbols: A Cultural History of Numbers*, trans. Paul Broneer (New York, 1969), pp. 263–5; and David Diringer, *The Story of the Aleph Beth* (New York, 1960), pp. 179–80.

letters, and the number of the beast, 666.[13] As Firth notes, such a table is not unique to Foxe's apocalypse commentary, but it does hint that his silence on the subject of the numerological significance of 294 is not the result of mere ignorance. His invented number, arguably, comes to have not only the exoteric meaning of a countdown to the end of the church's persecution, but also an esoteric meaning to be discovered by more thorough, wide-ranging, and astute readers. Foxe's discussion of a calculation explicitly forbidden—appointing the end of persecution and the Second Coming of Christ—asks readers to imagine the world as possessing a pattern that is "discoverable" by the transparent manipulation and invention of numbers.

Foxe insists that the autobiographical account of his oneiric inspiration for multiplying 42 and seven, as well as his verification of the calculation with third parties, should demonstrate to readers that he does not "follow any private interpretation of mine own."[14] Yet his narrative avoids the readily available numerological evidence that would most concisely serve to buttress this defense. In short, this admittedly brief episode in Foxe's massive volume—but one that is central to the organizational plan and historical theology of the entire martyrological project—does not present discovery as the ground or norm from which invention deviates. If anything, this personal narrative treats individual inspiration, and the imaginative inventions that flow from it, as an entirely sufficient ground for determining historical periodization, if not the generally conceived "truth of things."

This silent use of Kabalistic numerology is not a one-off event in the *Actes and Monuments*. The Rose Allin episode, "Tenne Martyrs condemned and burned at Colchester," closes with an arithmetical tally of the ages of the 10 martyrs, including Allin, who die at the stake: "Thus ended all these glorious x. soules that day, their happy liues vnto the Lord, whose ages all did growe to þe summe of 406. yeares or thereabouts."[15] This discovery is apparently important enough to be repeated in a marginal gloss: "The age of these Tenne made the summe of 406." The book's brief obsessive repetition of this number seems completely inexplicable until we recognize the numerological value of 406: it is the value of *atah*, or "thou." Or it is the numerical equivalent of *tav*, the last letter of the Hebrew alphabet, spelled out (i.e., since *tav* = 400 and *vav* = six, when one spells out *tav—tav-vav*—the sum that results is 406).[16] It is probably this latter significance that Foxe's evocation is designed to produce, although the absence of commentary in the *Actes and Monuments* on this score makes such claims necessarily speculative. At the very least, if 406 signifies nothing more than this last letter of the Hebrew alphabet, it would highlight not only the finality of this martyrdom, but, simultaneously,

[13] Firth, *The Apocalyptic Tradition*, p. 98; John Foxe, *Eicasmi seu meditationes in sacram Apocalypsin* (London, 1587), p. 270.
[14] Foxe, *Actes*, bk 1, p. 100.
[15] *Ibid.*, bk 12, p. 2009.
[16] *Godwin's Cabalistic Encyclopedia*, p. 557.

its status as something more than a mere terminus—a significant concluding event with a meaning. Yet as Chanita Goodblatt explains, *tav* possesses such a variety of meanings that even this relatively full explanation does not exhaust its potential numerological significance. One of these meanings stems from the original orthography of the letter itself: "the original form of the letter 'tav' is a cross turned on its side." Even more likely in the context of the Allin episode is the meaning imagined by Calvin and John Donne, both of whom think it is the "protective marking of the righteous against their destroyers" mentioned in Ezekiel 9:4.[17]

As we saw with 294, in this instance the full weight of 406 is the result of extensive Kabalistic and semiotic study; yet the number itself is the apparently arbitrary invention of Foxe's imagination. The only difference is that this time it's not an oneiric vision that provides the invented number, but rather Foxe's careful observation and summation of the martyrs's ages. In noting this difference, I do not wish to suggest that Foxe's book contradicts itself or overcompensates, showing him as arithmetically inept in Book 1, only to present him as mathematically astute in Book 12. Rather, what matters here is how these different numerical inventions collude with numerology to produce a conflation of discovery and invention; and a commitment to, if not faith in, a very specific type of historical and even ontological pattern, one that is immanently and immediately knowable and present in texts.

Foxe's evocation of 406 in the Rose Allin episode is not just a means of tempting readers to interpret and decode numerological signs, of postulating an esoteric mystery that an initiated elite might discover. By dotting the page with this number, Foxe's text also bows to the imperative of its encyclopedic recording apparatus. Thus, it is not just a matter of discovering what 406 means, but rather of recording and repeating 406, of reaffirming the invention and discovery of this number and its importance for a broader pattern of available truth. After all, this number does not just appear briefly in the text and disappear as a minor detail: the marginal gloss reaffirms its mysterious importance—"The age of these Tenne made the summe of 406."[18] 294 receives similar treatment, but to a much greater extent, in the *Actes and Monuments*. One table, "The mysticall numbers in the Apocalyps opened," explains the interrelations between the pivotal numbers from the Book of Revelations—42 months, 1260 days, three and a half years, and three and a half days—but ultimately concludes by reaffirming the equation of 42 months and 294 years.[19] The more important table, though, "A Table containing the time of the persecution both of the primitive, and of the latter Church, with the count of yeares from the first binding up of Sathan, to his loosing againe, after the minde of

[17]   Chanita Goodblatt, "From 'Tav' to the Cross: John Donne's Protestant Exegesis and Polemics," in Mary Arshagouni Papazian (ed.), *John Donne and the Protestant Reformation: New Perspectives* (Detroit, 2003), pp. 221–46; pp. 234–5.

[18]   Foxe, *Actes*, bk 12, p. 2009.

[19]   *Ibid.*, bk 1, p. 101.

the Apocalyps," appears intent on populating the page with the numbers 294 and 1294.[20] In other words, the table is less interested in recounting an accurate history than in discovering, producing, or recording the number 294. Moreover, Foxe's book has increasingly obliterated the distinctions between these activities.

Although these admittedly brief moments in the text may seem mere curiosities in the otherwise familiar and transparent propagandistic plan of the *Book of Martyrs*, the use of number and numerology in these instances also reveals Foxe's broader understanding of the relationship between encyclopedic recording and events. As we will see, the approach here is nothing so banal as "recording makes it so," or the sheer performativity of inclusion and exclusion. Rather, the impulse of the *Actes and Monuments* to totality must promise, even if it cannot deliver, a world and a text in which the dialectic between inclusion and exclusion—and, for that matter, invention and discovery—no longer operates.

"An Universall History of the Same"

Foxe's compendium poses a special sort of problem because it is not just any old Protestant encyclopedia, but rather an attempt, in Breitenberg's phrase, "to amass an entire Protestant 'state' of texts."[21] That is, the martyrological project demands that this total book not just represent events, or use them to debate the merits of various confessional allegiances; but also that it perform the sort of immanent manifestation of true faith embedded in the rhetorical notion of *martyria*. Thus while Foxe's encyclopedic impulses and chronicling procedures are certainly not unique, his project's attempt to produce a totally recorded world, where mystery is not discovered and yet the martyr's meaning is not retroactively invented, poses a basic problem for any readership: viz., what does the *Actes and Monuments* ask of us, if it removes both suspense and retroactive interpretation from our repertoire of activities?

Of course, the answer to these questions revolves, at least in part, around the burgeoning size and ambition of the *Actes and Monuments* over Foxe's lifetime, its transformation from a history with very specific types of evidence to a "universal history." Even the title changes register this propensity. The 1563 edition promises a very narrow recent history with a particular focus on England and Scotland:

> Actes and Monuments of these latter and perilous dayes, touching matters of the Church, wherein are comprehended and described the great persecutions & horrible troubles, that have bene wrought and practiced by the Romishe Prelates, speciallye in this realme of England and Scotlande, from the yeare of our Lorde a thousande, unto the tyme nowe present.

[20] *Ibid.*, bk 5, pp. 397–8.
[21] Breitenberg, "The Flesh Made Word," p. 392. For another account of the book's impulse to totality, see Warren Wooden, *John Foxe* (Boston, 1983), pp. 41–2.

By 1583, the book has expanded to include the early persecutions, but it is not just this historical extension that differentiates the final edition. As several critics have noted, not only does the 1583 title dispense with a description of the types of evidence used to support its historical claims; it also arrogates to itself the project of a "universal history," one in which there is not so much a special focus on England and Scotland, as there is an identical repetition of earlier persecutions:[22]

> Actes and Monuments of matters most speciall and memorable, happenying in the Church, with an Universall history of the same, wherein is set forth at large the whole race and course of the Church, from the primitive age to these latter tymes of ours, with the bloudy times, horrible troubles, and great persecutions agaynst the true Martyrs of Christ, sought and wrought as well by Heathen Emperours, as nowe lately practiced by Romish Prelates, especially in this Realme of England and Scotland.

Yet what would it mean to write, much less read, "an universal history" of the church? Would not "an *Universall history of the same*" not be a history, strictly speaking, at all?

There is more to Foxe's copiousness than the mere accumulation of ever more evidence, documentation, and repetitive stories. At some point, the very nature of the martyrology changes, and it seeks to accomplish something other than the mediated representation of events. As Breitenberg notes:

> By a variety of narrative strategies, Foxe seeks to collapse the inevitable mediation between the event and its textual depiction in the *Acts and Monuments*. This desire to reproduce the original event imitates the larger design of Foxe's history, which is, in part to reproduce the original purity of the apostolic church before its corruption by Rome ... Reproductions of the Word, the apostolic church and, in the case of the scene at the stake, the crucifixion, consequently possess a great investment in understanding their representations as unmediated reproductions. In the case of the Word, language is understood as plain and open rather than allegorical; as for Foxe's depictions of martyrdom, the strategy is that his own text fully reproduces the event.[23]

Yet the rhetorical aim that Breitenberg ascribes to Foxe's "reproductions" reintroduces mediation at precisely the point where the text most stridently seeks to overcome it: after all, an event reproduced in a text has its primary existence somewhere else. Foxe's text does not subordinate its impulses toward immanent

---

[22] For the argument that the editions published in Foxe's lifetime progress from prophetic to apocalyptic to monumental history, see Tom Betteridge, "From Prophetic to Apocalyptic: John Foxe and the Writing of History," in Loades (ed.), *John Foxe and the English Reformation*, pp. 210–32; and Woolf, "The Rhetoric of Martyrdom," p. 245.

[23] Breitenberg, "The Flesh Made Word," p. 396.

inclusiveness, an impulse embodied both in the notion of *martyria* and the goal of a Protestant encyclopedia, to some more reputable end; but rather is unapologetic in its affirmation of the exclusively textual reality of all of its events. The *Actes and Monuments* is not a conglomeration of events outside in the world, discovered or invented, but rather precisely what its title maintains: not just a monument to these events, but the acts that constitute these events themselves. In short, Foxe's book fancies itself both the story and the meaning, the sign and its significance.

The function of repetition and encyclopedic inclusiveness in Foxe is to assure us that there is only sameness in the world: there are no irreconcilable debates, differences, or others. Narrative cannot accomplish this, insofar as it reaffirms the basic difference between story and meaning, between past and present and future. In contrast, number, particularly chronology, appears as a pivotal, if multifarious, aspect of Foxe's drive toward homology. That is, unlike the numerological temptations that this chapter has already addressed, chronological numbers promise to manifest sameness without resorting to a necessarily self-undermining or absent meaning. *Chronos* and chronology then promise what numerology cannot: a notion of manifestation that does not depend on a secret and retroactive interpretative process. As even the organization of Foxe's text demonstrates, the explanation of 294 requires not only an esoteric intertextuality, but also, quite simply, delay, in this case a delay of 100 pages. Numerology depends upon a hermeneutic practice that distributes meaning after the fact, even if it ultimately asserts that it's always been there, all along, inside of the numbers. A meaningful *chronos* appears attractive precisely because it holds out the fantasy of an immanent fullness, the witness and testimony that *martyria* demands.

As Alison Chapman notes, Foxe's Kalendar seems so obsessed with maintaining chronological sequence that the accuracy of death dates bow to this imperative.[24] Yet this is not just the function of the introductory Kalendar, but also a result of an episodic chronicle structure that presents the date of death, the sheer terminus of a martyr's life, as the principle of organization for the entire volume. As John King notes, instead of unfolding through a full chronological sequence, all narrative material appears appended to the date of martyrdom.[25] By arranging the individual martyrs's lives according to the dates of death, rather than a chronological narration of their participation in Reformation debates or events, the volume's organization actually implies that it is neither the death nor the cause that makes the martyr, but the temporal sequence itself.

The typographical arrangement of the final three books of the *Actes and Monuments*, which treat the Marian persecutions, reaffirms the pivotal role of numbers in organizing the volume. Printed in the left hand margin of most pages

---

[24]   Alison A. Chapman, "Now and Then: Sequencing the Sacred in Two Protestant Calendars," *Journal of Medieval and Early Modern Studies*, 33 (Winter 2003): pp. 92–123; p. 96.

[25]   John N. King, *Foxe's Book of Martyrs and Early Modern Print Culture* (Cambridge, 2006), p. 249.

is the date of the events chronicled in the text, presumably as a finding aid in this gargantuan tome. These notations occur all the way up to page 2121, three pages before Foxe transitions into his appendices. Earlier books of the *Actes* have similar chronological finding aids, but these appear more sporadically. The dates involved frequently do not match, as a result of the volume's organization of documents and narratives around death dates. Thus, as just one example, "Anno. 1555 February" appears in the margin of an account of one of Hooper's examinations that occurs in January 1555. "THe xxij. of Ianuary followyng, 1555" is the first phrase on the page, right beside the incorrect marginal notation.[26] Even more ill-fitting is Foxe's history of the invention of the Catholic mass: this polemic recitation of over 1200 years of liturgical history occurs on pages marked "Anno 1553."[27] Dates even appear on pages where there are no events to date. Thus, "Anno 1555" appears in the margin of Book 10's reproduction and refutation of the mass.[28] In sum: always haunting this text from the margin, regardless of what's in the text itself, is a reaffirmation of an organizational system based on the mere march of time— *chronos*—not what happens in it.

A notation that looks to be nothing more than an indexing and cataloguing instrument offers an order that competes with the narratives contained in the book. Even the conclusion of the account of a pivotal martyr like Thomas Cranmer cannot avoid mentioning his mere temporal position in the broader historical drift of the Marian persecutions:

> And thus haue you the full story concernyng the lyfe and death of this reuerend Archbish. and Martyr of God, Thomas Cranmer, and also of diuers other the learned sort of Christs Martyrs burned in Queene Maries time, of whom this Archb. was the last, beyng burnt about the very middle tyme of the raign of that Queene, and almost the very middle man of all the Martyrs which were burned in all her raigne besides.[29]

"Almost the very middle man." The passage does not insist that this approximate median status means anything: it merely places Cranmer in a chronological sequence. Readers might certainly conclude that the detail has significance, that being in the middle means being the fulcrum or pivot of events. Yet this is to read story, drama, and climax into a book that concludes its account of Cranmer not with his dramatic death, but with his letters.

---

26 Foxe, *Actes*, bk 11, p. 1507.
27 *Ibid.*, bk 10, pp. 1404–5.
28 *Ibid.*, bk 10, p. 1399.
29 *Ibid.*, bk 11, p. 1888.

Conclusion

The point of this admittedly hasty tour through the manifold ways that Foxe deploys chronology is not simply that *chronos* (the sheer march of numbers) rather than *kairos* (a special opportunity seized) organizes the book.[30] Rather, what matters for us here is what sort of world *chronos* produces for readers: like 294 and 406, a sequence of dates is not really discovered, but neither is it a sheer act of invention. Chapman describes Foxe's Kalendar and the version of time it embodies as "a relentlessly sequential unfolding of historical time." For Hayden White, the chronological unfolding of annals is similarly "relentless": "There is no scarcity of the years: they descend regularly from their origin, the year of the Incarnation, and roll relentlessly on to their potential end, the Last Judgment."[31] However, framing Foxe's use of numbers and chronology as an inexorable march assumes the very sort of dialectical fight between determinism and human agency—and discovery and invention, for that matter—that Foxe's encyclopedic inclusiveness seeks to evade: i.e., the march of time is only relentless if one fears the Second Coming or if one imagines such a march as necessarily at odds with one's own desires.

Andrew Escobedo's account of the conflict between apocalyptic and national histories nicely encapsulates the need for a non-dialectical account of invention and discovery if one is ever going to imagine providential history as something more than a fatalistic determinism. "Modernity, like progress," Escobedo writes, "requires an effort of self-definition (an 'accomplishment') that dramatically, if unrealistically, leaves the past behind. Human effort thus seems to produce a break in time, contrary to the notion of historical progress as inevitable, gradual, and continuous."[32] Foxe's book, as I have attempted to show throughout this chapter, does not accept this basic dialectic, but rather attempts to render the world both flatly and immanently homologous: no dramatic eruptions of accomplishment or inevitable bows to fatalistic forces here. Instead, Foxe's text presents events as something other than ruptures in the immanent order of the world that then require interpretation in order to make them fit sensibly within the briefly disordered whole. And that, again, is what the Protestant encyclopedia's total inclusiveness attempts: a notion of invention that does not react against discovery; one that is not bound forever to the dialectical interchange between the given and the agentially created.

Number and chronology have a pivotal function in such an account. Unlike kairotic opportunities, chronology is neither invented nor discovered, and offers no opportunities for tension, dialectical or otherwise. There is no problem of

---

[30]   See Hayden White, "The Value of Narrativity in the Representation of Reality," in *The Content of the Form: Narrative Discourse and Historical Representation* (Baltimore, 1987), p. 8.

[31]   Chapman, "Now and Then," p. 96; White, "The Value," p. 11.

[32]   Andrew Escobedo, *Nationalism and Historical Loss in Renaissance England: Foxe, Dee, Spenser, Milton* (Ithaca, 2004), p. 209.

summing up, or subordinating events to a broader teleological end. The turn to *chronos* does not simply reject meaning, but rather transforms what we mean by signification and meaning. The notion that one would uncover a secreted meaning in the simple march of time is belied by our earlier discussion of numerology: the significant numbers are not the object of discovery, but rather invention. Yet neither does the *Actes and Monuments* unleash the capricious desires of readers, providing them with the freedom to invent at will. What both of these gestures share is the notion that meaning resides somewhere else, in the reality or secret unearthed, or in the interior recesses of the inventing subject, or the beneficial consequences of the invention. There is certainly that aspect to martyrdom, the uncovering of the true cause; but it ignores, as Foxe's use of numbers repeatedly emphasizes, the notion of a readily apparent and available manifestation. Number, both in its numerological and chronological capacities, becomes the engine for overcoming this particular conundrum within Foxe's martyrology.

Chapter 9
Unearthing Radical Reform: Antiquarianism against Discovery

Travis DeCook

In *Hydriotaphia* (1658), Thomas Browne takes the discovery of ancient funeral urns in Norwich as an occasion to meditate on the diversity of burial customs across a wide array of historical periods and cultures. Despite the text's bravura display of antiquarian learning, and the excitement it conveys about the discovery of these ancient and bizarre artifacts, scholars have noted that, for Browne, unearthing the past is a matter of confusion, loss, and incompleteness.[1] In fact, the text seems to undermine the very promise of discovery: throughout *Hydriotaphia*, we are confronted with the yawning gulf between present and past, and the impossibility of complete historical knowledge. By the text's apocalyptic end, the hope of immortality through commemoration is rendered pointless in the face of eternity—as is, implicitly, the antiquary's drive to discover.[2]

Thus though antiquarianism is undoubtedly one of the fields of early-modern knowledge to which discovery is foundational, Browne's text can be situated within a countervailing tradition of "anti-discovery." Writings that undermine discovery in a similar way include Joseph Hall's satire *Mundus Alter et Idem* [*A World Different and the Same*]—a fictional account of travels to new lands, paradoxically discovering nothing but the same folly and vice rampant in Europe—and several of the poems in George Herbert's *The Temple*. For example, "Vanity (1)" describes the discoveries of a diver, chemist, and astronomer, but sharply undercuts them in the final lines:

> What hath not man sought out and found,
> But his dear God? who yet his glorious law
> Embosoms in us, mellowing the ground
> With show'rs and frosts, with love and awe,
> So that we need not say, Where's this command?
> Poor man, thou searchest round

---

[1]   See Claire Preston, *Thomas Browne and the Writing of Early Modern Science* (Cambridge, 2005), p. 132; and Philip Schwyzer, *Archaeologies of English Renaissance Literature* (Oxford, 2007), p. 180.

[2]   Preston, *Thomas Browne*, p. 150.

To find out *death*, but missest *life* at hand.[3]

Rather than recognize the manifest presence of God in the world around us, people seek out hidden but unprofitable knowledge. For Herbert, this is inevitably debased and carnal, because experienced without relation to the divine. Discovery, in short, is dangerous; and it is dangerous because it is a missing of the point.

This chapter considers the ways in which the clergyman Thomas Fuller, one of the seventeenth century's greatest historians, contributes to the period tradition of anti-discovery. For Fuller, as for Browne, a process of discovery stands in tension with the perspective of eternity. Yet unlike Browne, Hall, or Herbert—who criticize specific acts or modes of discovery—Fuller chooses metaphors of discovery to attack specific, and opposed, errors of historical engagement. He attacks the evacuation from history of sacred significance, on the one hand, and the fetishization of the antiquarian past, on the other. Much has been made of Fuller's objection to extremism: he wrote an essay titled "Of Moderation" addressing destructive factions,[4] and indeed moderation is a recurring touchstone in virtually all of his writings, reflecting his commitment to the established church, and his rejection of both Laudian and Puritan excesses.[5] Central here is what Florence Sandler recognized as Fuller's emphasis on the church's historical nature—its continuity and situatedness in time—against extremist emphases on temporal discontinuity and ahistorical perfection.[6] It is this vision that motivated Fuller's antiquarian projects *A Pisgah-Sight of Palestine* (1650) and *The Church History of Britain* (1655), which, respectively, provide "the appreciative chronicling of God's dealings with the historical Israel, and then with the historical New Israel, the Church."[7]

The fraught status of history during this revolutionary age is articulated by Fuller through elaborate metaphors of discovery. Unsurprisingly, discovery is not always portrayed negatively by Fuller: for example, in *The Holy State* (1642), Fuller praises the antiquary William Camden's discoveries of England's past, which have enabled new forms of national self-knowledge.[8] However, in key places, metaphors of discovery are deployed by Fuller in order to castigate the

---

[3]   George Herbert, "Vanity (1)," in John Tobin (ed.), *George Herbert: The Complete English Poems* (London, 1991), p. 79, ll. 22–8.

[4]   Thomas Fuller, *The Holy State* (London, 1642), pp. 205–8.

[5]   For the seventeenth-century "discourse of moderation" and its ideological deployment, see Robert M. Calhoon, "On Political Moderation," *Journal of the Historical Society*, 6/2 (2006): pp. 275–95; 282.

[6]   Florence Sandler, "Thomas Fuller's *Pisgah-Sight of Palestine* as a Comment on the Politics of Its Time," *Huntington Library Quarterly*, 41/4 (1978): pp. 317–43; 326.

[7]   Florence Sandler, "The Temple of Zerubbabel: A Pattern for Reformation in Thomas Fuller's *Pisgah-Sight* and *Church-History of Britain*," *Studies in the Literary Imagination*, 10/2 (1977): pp. 29–42; 33.

[8]   Fuller, *Holy State*, pp. 146–7.

views of history that he associates with religious radicals, of one kind or another. Fuller's commitment to moderation can be understood in terms of his affirmation of the church's mediating functions within history; his support for the established church and its gradual rather than radical reformation entails a traditional view of the church as historically embedded, as against a radical belief in discontinuity and rupture.

At the same time, as an antiquary, Fuller is sensitive to the capacity for undue attention to the historical past. In traversing the Scylla of a radical, millenarian flattening of history, and the Charybdis of an idolatrous privileging of it, Fuller invokes discovery to critique presumed ways to knowledge and understanding of the relationship between knowledge and history itself. This chapter will first address Fuller's polemic against the radical rejection of history in *A Sermon of Reformation* (1643), and will then turn to his acknowledgement, in *A Pisgah-Sight of Palestine*, of the idolatrous valorization of historical knowledge, at the expense of religious imperatives.[9]

Discovery and Reform

Delivered during a time of chaotic sectarianism, *A Sermon of Reformation* encourages gradual improvement against the desire for radical change. The latter is presented as unrealistic, and resulting from a blindness to history expressed in a narcissistic privileging of the contemporary moment. This privileging enables the radicals to believe they are founding something new rather than repairing something old, acting as if they are the first Christians to appear in England and assuming others are "pure Pagans where the word is newly to be planted."[10] Fuller recounts a narrative in which a Spanish duke's followers, upon searching for a lost hawk, discover a "new Country in the Navell of Spaine, not knowne before, invironed with Mountaines, and peopled with naked Salvages."[11] The satire hinges on the lunacy of believing that such "a Terra Incognita could be found in England; which (what betwixt the covetousnesse of Landlords and the carefulnesse of Tenants) is almost measured to an Acre." Fuller goes on to acknowledge that if in England "such a place were discovered, I must allow that the Preachers there were the first planters of the Gospel, which in all others places of the kingdom are but the Continuers thereof."[12] The impossibility of such a *terra incognita* within England,

---

[9]   For the early-modern understanding of idolatry as valuing people and objects for their own sake, rather than referring them to their higher ends and created status, see David Hawkes, *Idols of the Marketplace: Idolatry and Commodity Fetishism in English Literature, 1580–1680* (London, 2001), pp. 3–5.

[10]   Fuller, *A Sermon of Reformation* (London, 1643), p. 20.

[11]   *Ibid.* The story is taken from James Howell, *Instructions for forreine travel* (London, 1642), p. 134.

[12]   Fuller, *A Sermon*, p. 20.

however, points to the equally impossible notion of an England untouched by history. For Fuller, his opponents' fascination with discovery is predicated on a failure to grasp the church's historical nature.

Significantly, in his refutation of Fuller, the radical preacher John Saltmarsh invokes the trope of discovery to illustrate the possibility of perfecting the church. "He that looks abroad," Saltmarsh writes, "shall soone have his sight terminated but the more he goes on, the more hee sees, and that which closed his prospect, opens then into new discoveries; if you see no perfect Reformation as you stand, doe not therefore say there is none."[13] In such a radicalized critique of the established church, show of worship is a mere appearance which, when penetrated, reveals the reality of an absence of true religion. Fuller, however, answers with his own allegory of discovery. Like Saltmarsh's image of ever-increasing vistas, this emphasizes gradual process, but without positing a final or terminal state:

> Mariners which make forth for the Northerne Discoveries, goe out with this assurance, that it is impossible to come to the pole. Yet have they sought and found out very farre, almost to the eightieth degree of latitude. What covetousnesse or curiosity did in them, sure Grace is as active to doe in Gods Children who will labour *to draw neere* to a perfect Reformation, in obedience to Gods command, though they know they shall never fully attaine unto it.[14]

Fuller's point is that while the church the radicals criticize may be imperfect, it is no less legitimate for that. Indeed, Fuller relegates his opponents' belief in a once-and-for-all reformation to utopia: "a perfect Reformation of any Church in this world may be desired, but not hoped for. Let Zenophons Cyrus bee King in Platos Common wealth; and Batchelors-wives breed maides children in Mores Utopia ... These phansies are pleasing and plausible, but the performance thereof unfeasable, and so is the perfect reformation of a Church in this world."[15]

Fuller's argument with Saltmarsh bears a structural similarity to one of the earliest polemical exchanges in the English Reformation: Thomas More's assault on Martin Luther. Echoing Augustine's conviction that the City of God can never be fully manifest on earth, More was committed to a church of saints and sinners: an imperfect but visible, firmly established institution, the only place within the fallen world for people to commune with God. Accordingly, More attacked Luther and his English follower William Tyndale for their doctrine of an invisible church of the elect, a congregation of saints not tied to the visible Roman church or any

---

[13] John Saltmarsh, *Examinations, or, a Discovery of some Dangerous Positions* (London, 1643), p. 9.

[14] Fuller, *Truth Maintained, or Positions Delivered in a Sermon at the Savoy* (London, 1643), p. 5.

[15] *Ibid.*, pp. 23–4.

particular place, but existing halfway between earth and heaven.[16] In a marginal note printed in More's *Responsio ad Lutherum*, Luther's notion of a church containing only true believers hidden in the midst of the actual, fallen world is mocked with the remark "perhaps he has seen this in utopia."[17] The belief in an invisible church of the elect is derogated not merely as the belief in perfection existing in the world, but also, through this allusion to More's own classic text of geographical discovery, as a delusion that the true church exists beyond appearances and requires an act of discovery. Instead, for More, the church must be understood as necessarily evident—always recognizable—lest we be confronted with total uncertainty.[18]

Discovery is a central predicate of the reformers' privileging of Scripture, and implies and reproduces the separations instituted by the Reformation. *Pace* J.G.A. Pocock's contention that the Reformation drew the process of salvation more directly into history,[19] Tyndale portrays divine revelation as unchanging and consequently *outside* history: contained in the Word of God, and encountered by the true church of the elect in a relatively stable form.[20] Tyndale's rhetoric emphasizes the true church's relative changelessness, even though it exists within and throughout historical time. This stasis is the mirror image of the transhistorical understanding of revelation rooted in *sola scriptura*, which emphasizes the completeness and finality of Scripture's record of revelation.[21] In contrast, More affirms the shifting, developing nature of the Roman Catholic Church as the forum for God's communication with humanity.[22] The historical church participates in the divine, for More, history being the very vehicle through which revelation occurs and unfolds.

---

[16]   See David Weil Baker, *Divulging Utopia: Radical Humanism in Sixteenth-Century England* (Amherst, 1999), pp. 65–6. See also Jaroslav Pelikan, *The Christian Tradition: A History of the Development of Doctrine* (Chicago, 1985), vol. 4, pp. 174, 271–2. Anticipating Fuller's reference to "Plato's Common-wealth," Luther in fact denied that his notion of the church was a "Platonic republic" (in Pelikan, p. 174).

[17]   Thomas More, *Responsio ad Lutherum*, ed. John M. Headley, in *The Complete Works of St. Thomas More* (15 vols, New Haven, 1963–97), eds Louis L. Martz, Richard S. Sylvester, and Clarence H. Miller, vol. 5, p. 119. Hereafter, this edition will be abbreviated *CW*.

[18]   *CW*, vol. 5, p. 167.

[19]   J.G.A. Pocock, "Time, History, and Eschatology in the Thought of Thomas Hobbes," in *Politics, Language and Time: Essays on Political Thought and History* (New York, 1971), p. 178.

[20]   See Tyndale, "A Prologue into the Second Book of Moses Called Exodus," in Thomas Russell (ed.), *The Works of the English Reformers: William Tyndale and John Frith* (London, 1831), vol. 1, p. 24; and *An Answere vnto Sir Thomas Mores Dialoge*, in Anne M. O'Donnell and Jared Wicks (eds), *The Independent Works of William Tyndale* (Washington, DC, 2000), vol. 3, p. 25.

[21]   Tyndale, *An Answere*, pp. 24–6 ; and *The Obedience of a Christian Man*, ed. David Daniell (London, 2000), pp. 17, 52.

[22]   *CW*, vol. 6, p. 146; and vol. 8, pp. 284, 337.

Unlike More's, Fuller's understanding of the church was not dogmatically exclusive, but was relatively accommodating and did not involve a strict divide between true church and heresy; rather, his focus was on negotiating the boundary between matters essential and matters indifferent in church governance and worship. An heir to *sola scriptura*, Fuller saw revelation as complete and contained in the Bible, rather than as an ongoing process.[23] Despite these points of difference, however, More and Fuller are obviously united in defending the continuity of an established church in the face of reformers affirming a source of revelation outside its bounds. For Tyndale, this is simply to be found in scripture itself; Saltmarsh goes much farther, asserting that God inspires select individuals through "secret *preparations*."[24] Yet both Tyndale and Saltmarsh emphasize that the truth *needs* to be discovered; that the true Word of God or authentic mode of worship is beyond the appearances of historical reality. History is separated from revelation, and the past becomes, at least implicitly, evacuated of truth and meaning.

Fuller's opposition to this idea can be framed by the conflict traced by Achsah Guibbory in the seventeenth-century English church: Puritans against "ceremonialists," who maintained historical continuity and the legitimacy of tradition. The ceremonialists defended tradition as subservient to Scripture yet nonetheless meaningful, positing a relation of interdependence, harmony, and hierarchy, opposed to the Puritans' uncompromising dualisms.[25] Since this position implies the participation of ceremony and tradition with the divine, rather than the stark binary of Scriptural/non-Scriptural, it is consonant with dominant strains of pre-modern, pre-reformed thought. Along these lines, ceremony was also defended on the universalist grounds that it accords with nature and reason, and that these exist in continuity with Scripture.[26] On the other hand, for the Puritan habits of thought discussed by Guibbory, history and creation are not to be construed in terms of mediation, continuity, and participation with the divine—that is, in the terms we have seen in Herbert's "Vanity (1)." Rather, the divine is *to be discovered*, within, but also apart from, history and creation. This is consistent with burgeoning late-medieval and early-modern understandings of God as absolutely other to creation. It has often been argued that such theological shifts set the stage for the construction of the autonomous history and nature of secular modernity.[27]

While Fuller opposed the ceremonial rigor of the Laudians, he shared their "Catholic" emphasis on continuity and integration, at least insofar as it served his political purposes. On the other hand, and despite his own rootedness in the reformed tradition, in his polemics Fuller drew on the secularity of discovery to attack the radicals' understanding of their historical role. He deployed a metaphor

---

[23]   Fuller, *Truth Maintained*, pp. 42–3.

[24]   Saltmarsh, *Examinations*, p. 7.

[25]   Achsah Guibbory, *Ceremony and Community from Herbert to Milton: Literature, Religion and Cultural Conflict in Seventeenth-Century England* (Cambridge, 1998), p. 29.

[26]   *Ibid.*, pp. 133–4.

[27]   Charles Taylor, *A Secular Age* (Cambridge, MA, 2007), pp. 773–4.

of discovery in order to attack religious radicals for rejecting the historical church and its mediating, temporal function.

History and Idolatry

In *A Pisgah-Sight of Palestine*, Fuller associates certainty in eschatological predictions with ignorance of history: "those, which seem to know all which is to come, know but little of what is past; as if they were the better Prophets, for being the worse Historians."[28] Ignorance of the "historical dispensation" is associated here with a fanatical and deluded self-assurance and conviction that one has received new revelations.[29] In this text, however, Fuller also connects the presumption of discovery, not to historical blindness, but to an idolatrous replacement of religion with history.

Taking its name from Moses' view of the Promised Land from the top of Mount Pisgah, *A Pisgah-Sight of Palestine* was described in its Stationer's Company entry as "a Choragraphicall Comment on the History of the Bible, or the description of Judaea."[30] Much of the text is organized according to the tribal lands of ancient Israel. This chorographical structure contains within it descriptions of the significant historical events and personages associated with each region. Moreover, the book incorporates discussions of antiquarian subject-matter, including a lengthy description of the Jerusalem Temple and the material culture of Temple worship; and discussions of the customs and religions of the ancient Biblical lands.[31]

At the opening of his text, Fuller raises three accusations that could be leveled against it: "that the description of [Palestine]: 1. Hath formerly been done by many. 2. Cannot perfectly be done by any. 3. If exactly done, is altogether uselesse, and may be somewhat superstitious."[32] Fuller's third remark connects superstition to undue attention to something useless. His reference to the uselessness of his work picks up on widespread criticism of antiquarian study as an utterly pointless enterprise.[33] But this uselessness takes on a specifically religious dimension here: "it matters not to any mans salvation, to know the accurate distance betwixt Jericho and Jerusalem; and he that hath climbed to the top of Mount Libanus, is not in respect of his soul, a haires breadth nearer to heaven."[34] Undue attention to such matters is construed as superstition.

---

[28]   Thomas Fuller, *A Pisgah-Sight of Palestine* (London, 1650), bk 5, p. 198.

[29]   Sandler, "The Temple of Zerubbabel," p. 33.

[30]   Quoted in Sandler, "Thomas Fuller's *Pisgah-Sight of Palestine*," p. 317.

[31]   Fuller, *Pisgah-Sight*, bk 4, p. 136.

[32]   *Ibid.*, bk. 1, p. 1.

[33]   See Daniel Woolf, *The Social Circulation of the Past: English Historical Culture 1500–1730* (Oxford, 2003), p. 175.

[34]   Fuller, *Pisgah-Sight*, bk 1, p. 2.

It is no surprise then that throughout the text Fuller is at pains to defend himself against the charge of giving undue attention to Biblical history, geography, and culture at the expense of the religion which gives these phenomena their ultimate meaning. There are two crucial contexts relevant to Fuller's defense of his work: the problematic nature of the Holy Land in Protestant thought, defined by an anxiety about ascribing sacramental power to place; and the vexed status of the study of antiquities, characterized by an entrancement with the artifacts of the past, at the expense of a proper orientation of one's spiritual energies to the worship of God.

In terms of the former problem, Fuller employs spiritual and allegorical interpretations of the Biblical lands: for instance, the relative absence of gold in the Holy Land is construed as God's emphasis on spiritual rather than worldly riches. However, his book, like many other post-Reformation historical studies of the Biblical lands and accounts of travel to the Levant, carefully distinguishes itself from the tradition of pilgrimage writings characteristic of the Catholic tradition, in which the tombs and relics of the Holy Land were imbued with sacramental power.[35] Indeed, Fuller perhaps most forcefully demonstrates his self-consciousness about this tradition by refusing to use the term "Holy Land" at all, for fear of suggesting a superstitious notion of the land's spiritual efficacy.[36]

The superstitious potential of the reception of antiquarian study was a similarly troubling prospect for Fuller. Despite the frequent uses of antiquarian scholarship throughout the sixteenth and seventeenth centuries to promote arguments for church reform, in the mid-seventeenth century antiquaries were generally identified with the established church and its traditions, and several antiquarian works were written in support of episcopacy.[37] Additionally, the period witnessed a long-running suspicion that antiquarian scholarship had links to popery, both because many antiquaries studied remnants of the Catholic past such as monasteries, and because the exhumed bodily remains studied by antiquaries could resemble saints' relics, particularly given the kind of fascination they conjured.[38] Additionally, the material culture sought out by antiquaries emphasized a visual orientation that raised iconophobic suspicions.[39] Owing to the artifact's capacity to enchant, and the resemblance of such enchantment to idolatrous devotion focused on material objects, the antiquarian past could be viewed as idolatrous.[40]

---

[35]   For this Catholic tradition, see Jaś Elsner and Joan-Pau Rubiés, "Introduction," in Jaś Elsner and Joan-Pau Rubiés (eds), *Voyages and Visions: Towards a Cultural History of Travel* (London, 1999), pp. 16–18; and Robert L. Wilken, *The Land Called Holy: Palestine in Christian History and Thought* (New Haven, 1994), pp. 115–22.

[36]   Fuller, *Pisgah-Sight*, bk 1, p. 4.

[37]   Graham Parry, *The Trophies of Time: English Antiquarians of the Seventeenth Century* (Oxford, 1995), pp. 18–19.

[38]   Woolf, *The Social Circulation*, p. 192. Fuller notes that the "aspersion" of loving popery "be generall on Antiquaries" (*Holy State*, p. 148).

[39]   Woolf, *The Social Circulation*, pp. 185, 189.

[40]   *Ibid.*, p. 148.

Even more broadly, in the charged atmosphere of the mid-seventeenth century, a militant ideology of purity meant that adherence to the past itself could be suspect, and was often attacked using imagery of antiquarian discovery. In Guibbory's words, "past human history, characterized by 'carnal traditions' and preserved through material, idolatrous 'monuments'" constituted a chief target for Puritans.[41] For instance, Peter Smart connected the external modes of worship upheld by the Laudian faction to a revived Jewish ceremonialism, and described this linkage using a metaphor suggestive of both antiquarian discovery and necromantic resurrection. For Smart, ceremonialism represents the exhumation of idolatrous error: "Jewish types and fygurs," he writes, were "long since dead and buried": to "revive" and "rayse" them up again is "an apostasy."[42] Crucial here is the issue of typology: the "Jewish types and figures," often in the form of external ceremony and material objects of Jewish worship, are to be valued only insofar as they foreshadow the superseding spiritual modes of Christian worship, a point made explicitly in *Pisgah-Sight*.[43]

A similar conjunction of idolatry, material culture, and undue reverence for the past can be found in John Milton's pamphlet *Of Reformation* (1641). Milton singles out William Camden, aligning the latter's support for bishops with his love of antiquities: "a fast friend of Episcopacie, *Camden*, who cannot but love Bishops, as well as old coins, and his much lamented Monasteries for antiquities sake."[44] Camden's is a compulsive, irrational love of anything old, characteristic of the "votarists of Antiquity."[45] Milton's critique combines a number of key issues. For one, it levels an attack on the idolatrous orientation towards the material past, figured by ancient coins which evoke the "guegaw's fetcht from *Arons* old wardrope" by the prelatical ceremonialists.[46] It also draws on the association between antiquarian study and the established church, as well as the problem of valuing the past for "antiquities sake." This latter issue reflects the illicit ascription of autonomy to the objects of historical scholarship. Rather than being understood and valued in relation to the imperatives of right religion, they become valued for their own sake, resulting in a kind of secular idolatry.

These associations between antiquarian study and an idolatrous veneration of the past are taken up by Fuller in *The Holy State*: "Some scoure off the rust of old inscriptions into their own souls, cankering themselves with superstition ... and they more lament the ruine of Monasteryes, then the decay and ruine of Monks lives."[47] Fuller notes that studying ancient artifacts can cause one to absorb the superstitions associated with them, and then acknowledges that artifacts can be

---

41   Guibbory, *Ceremony and Community*, p. 30.
42   Quoted in *ibid.*, p. 58.
43   Fuller, *Pisgah-Sight*, bk 3, p. 438.
44   John Milton, *Of Reformation* (London, 1641), bk 1, p. 11.
45   *Ibid.*, bk 1, p. 16.
46   *Ibid.*, bk 1, p. 3.
47   Fuller, *The Holy State*, p. 69.

fetishized, valued for their own sake: the monastery becomes of more concern than the monk it housed and the monk's beliefs. However, Fuller goes on to assert that a proper approach to antiquarian study only confirms one in correct religion, since it allows one to bypass manmade traditions and superstitions and gain an understanding of the primitive church.[48]

The idolatrous potential of antiquarianism is taken up again in Fuller's introduction to *Pisgah-Sight*, albeit with greater immediacy, since here he is concerned with deflecting these charges from his own work. Fuller's defensive introduction rests on a tension between historical and geographical knowledge and matters of religion, encapsulated in his statement about the irrelevance of the distance between Jericho and Jerusalem to one's salvation. What is significant here is that Fuller does not say, as we might expect, that if one goes on pilgrimage to Jerusalem or prays at a shrine in the Holy Land one does not get closer to heaven. Rather, he points out the uselessness of geographical measuring (the distance between Jericho and Jerusalem), and topographical experience (climbing Mount Libanus). In other words, Fuller homes in not on the idolatry of pre-reformed sacred geography, but on the superstitious adherence to proto-modern historical and geographical knowledge.

In the latter case, the relics of the past inspire fascination and veneration not because of any belief about sacral qualities imbuing material reality, but rather from a kind of inherent interest—a fascination with the ancient and alien for its own sake. Fuller's anxiety, I suggest, is about an emergent modern form of idolatry: it rests on a notion of historical distance, that is, on the pastness of the past, and it involves a secularized relic-worship involving a material culture construed as autonomous, outside of any regard for salvation or its relation to the divine. As Fuller puts it, "because the *New Ierusalem* is now daily expected to come down … these corporall (not to say carnall) studies of this terrestriall *Canaan*, begin to grow out of fashion, with the more knowing sort of Christians."[49] While an idolatrous approach to the Holy Land had been generally construed as the replacement of a properly spiritual religion with a carnal one, in this case a properly eschatological orientation is replaced with an unnecessary and distracting devotion to knowledge about material reality.

Prophecy and Recognition

One of the ways Fuller attempts to exculpate himself from the charge of an idolatrous veneration of the past is through an elaborate Biblical allusion to King Saul's forbidden conjuring of the prophet Samuel's spirit by the woman of Endor, described in 1 Samuel 28. Once again, Fuller connects a problematic attitude to the past with discovery, since this allusion hinges on the idea of an illicit (but

---

[48]   *Ibid.*

[49]   Fuller, *Pisgah-Sight*, bk 1, p. 3.

foiled) attempt at uncovering secret knowledge. Deployed in the specific context of Fuller's polemic, this attempt is suggestive both of an idolatrous reliance on another spiritual power than God, and of the presumption of human autonomy in relation to the divine.

In his response to the charge that his work is useless and superstitious, Fuller writes: "Besides, some conceive they heare *Palestine* saying unto them, as *Samuel* to *Saul* endevouring to raise him from his grave, *Why hast thou disquieted me to bring me up?*"[50] The raising of Samuel from his grave connects Fuller's project to the unearthing work of the antiquaries. Palestine is imagined as entombed bodily remains, one of the primary objects of antiquarian discovery and knowledge. This portrayal of Fuller's project as one of discovery jars with the work's title: Moses' view from Mount Pisgah, his initial sighting of the land promised to the Israelites, embodies the recognition of that which was promised rather than the discoverer's first glimpse of the new land. Along these lines, when Fuller acknowledges that much has already been written about the Biblical lands, he explicitly distinguishes his work from the ultimate event of discovery: "although we cannot with *Columbus*, finde out another world, and bring the first tydings of an unknown Continent or Island, by us discovered, yet our labours ought not to be condemned as unprofitable."[51] While prospective critics might construe his work of antiquarian and chorographical study as exhumation or necromancy, he explicitly distances it from discovery.

Attempting to exhume the remains of the Biblical past through an historical account, Fuller suggests, is unnecessarily disruptive: "Describing this Countrey is but disturbing it, it being better to let it sleep quietly, intombed in its owne ashes."[52] Beyond the charge of pointlessly expended effort, though, this fraught Biblical allusion implies that Fuller's work could represent a transgression of divine law. It is depicted as the idolatrous act of turning to sorcery rather than God, using necromancy to seek out knowledge. Fuller thus raises the possibility that studying the physical places and material culture of the Biblical lands represents a necromantic desire to speak with the dead. Such an evocation of religious transgression raises the contemporary association between the study of antiquities and idolatry.

A closer examination of the Biblical narrative that Fuller draws upon reveals the vital and complex role of discovery within it, and its relevance to the approach to history Fuller is exorcising. Confronted with the impending hordes of the Philistine army, Saul is anxious and wants to know what he should do. God does not answer his prayers for knowledge, however, because Saul has previously turned his back on God. Ironically, after he has outlawed the practice of necromancy in his kingdom, he turns to a woman known for her ability to conjure the dead.

---

50   *Ibid.*, bk 1, pp. 2–3.
51   *Ibid.*, bk 1, p. 2.
52   *Ibid.*, bk 1, p. 3.

In Fuller's text, this necromantic scene is mediated by the discourse of antiquarian scholarship, and Saul's divination figures Fuller's historical work as an illicit desire to discover, to unearth the past from its grave. What Saul obtains from Samuel, however, has a complex epistemological status that has bearing on the representation of discovery and historical knowledge in Fuller's text. First, if it is in fact a devil and not Samuel who is conjured, as many commentators (including Fuller) believed, the legitimacy of the "prophecy" delivered is obviously contestable.[53] According to Gregory of Nyssa, demons make signs pointing towards the future desired by those engaged in conjuring, so that conjuring results merely in the reflection of one's own desires rather than in discovery.[54] While the prophecy given to Saul does not reflect his desires, it undercuts discovery, since Samuel's words embody a full recognition of the situation rather than the revelation of something hidden. Samuel reiterates what Saul already knows: God is not responding because of Saul's previous rejection of the Lord. Samuel goes on to foretell Saul's and the Israelites' defeat on the basis of this rejection.[55] For Fuller, this is not prophecy. Instead it is a deduction based on what is open and evident. In *The Holy State*, Fuller asks how Satan could have known about Saul's imminent death, and suggests a kind of Satanic recognition brought about by reading the surface of things: having amassed 5,000 years of empirical data, Satan could "thereby make a more then probable guesse of future contingents; the rather because accidents in this world are not so much new as renewed."[56]

The possibility that the "prediction" of Saul's imminent demise is the result of recognition rather than discovery underlines Saul's own lack of understanding about the nature of his situation.[57] The king's conjuring assumes that it is possible to discover God's secrets. Faced with God's silence, Saul seeks out new knowledge on his own rather than coming to terms with the nature of and reasons for that silence. Indeed, his downfall can be understood as a failure to recognize the nature of his situation, which leads him to seek something not known. Saul's divination results from the temptation to discover, which replaces the moral requirement of recognition, of self-knowledge and acknowledgement of one's relationship to God. Saul's seeking necromantic knowledge results from his earlier failure to turn to

---

[53]   Fuller, *The Holy State*, p. 370.

[54]   Gregory of Nyssa, "Letter to Theodosius concerning the Belly-Myther," in Rowan A. Greer and Margaret M. Mitchell (trans), *The "Belly-Myther" of Endor: Interpretations of 1 Kingdoms 28 in the Early Church* (Atlanta, 2007), p. 171.

[55]   Gregory of Nyssa contended that the demon "uttered a likely deduction from what was apparent" as if it were a prophecy (*ibid.*, p. 173).

[56]   Fuller, *The Holy State*, p. 371. See also James VI, *Daemonlogie* (Edinburgh, 1597), pp. 4–5.

[57]   For the hermeneutics of recognition versus discovery, see Hans-Georg Gadamer, *Truth and Method*, 2nd rev. ed., trans. rev. Joel Weinsheimer and Donald G. Marshall (New York, 2004), pp. 110–19.

God for guidance and from his failure to read the clear signs of his own situation, namely, his alienation from God: there is literally nothing else to know but this.

In his commentary on this Biblical narrative, Joseph Hall homes in on the pointlessness of Saul's crime, calling it "bootless curiosity" and stating that knowledge about the future could not help him change its course. Saul's desire for necromantic knowledge is condemned by Hall as a superstitious mixture of the illicit and the useless, the "itch of impertinent and unprofitable knowledge" inherent in fallen humanity.[58] Similarly, as Fuller writes in *The Holy State*, "Mens minds are naturally ambitious to know things to come: Saul is restlesse to know the issue of the fight. Alas, what needed he to set his teeth on edge with the sourenesse of that bad tidings, who soon after was to have his belly full thereof."[59] Saul's conjuring thus dramatizes the seeking of knowledge that does not need to be sought; considered in terms of the paradigm of discovery, it is an ur-narrative of anti-discovery. The presumption to discovery in this instance is both deluded, in that there is nothing to discover, and idolatrous, in that it places the self in relation to knowledge in such a way that God is excluded; that is, it posits an epistemic relationship definitive of secular modernity.

From a Christian perspective, seeking knowledge from God necessarily involves the humble awareness that we can never know all God knows; that any knowledge we could ever achieve would always be incomplete. In stark contrast, Saul's transgressive necromancy represents the desire for knowledge as ego-centered mastery, excluding God, and trying to pry into secrets independently of one's contingent relationship to the divine. Fuller thus invokes a complex narrative in which knowledge is construed independently of the divine, God is replaced with self, and the very possibility of discovery is foiled. By raising these issues, Fuller presents himself as fully conscious of the threat of an idolatrous approach to scholarship and the historical past in order to neutralize such a charge against his own work.

Conclusion

Returning now to the tradition of "anti-discovery" with which this chapter began, Fuller can be linked to the conservatism of Browne and Hall. As Robert Appelbaum points out, Hall's dismantling of the possibility of discovery in *Mundus Alter et Idem* generates political quietism, his bleak *contemptus mundi* supporting acceptance of the *status quo*.[60] Philip Schwyzer has suggested that Browne's misattribution of the Norwich urns as Roman rather than Anglo-Saxon may be politically motivated.

---

[58]   Joseph Hall, *Contemplations on the Old Testament*, in *Bishop Hall's Works* (Oxford, 1837), vol. 1, p. 377.

[59]   Fuller, *The Holy State*, p. 369.

[60]   Robert Appelbaum, "Anti-geography," *Early Modern Literary Studies*, 4/2 (1998): par. 13.

Since this misattribution further alienates the present from the past, identifying as it does the discovered remains with foreign invaders rather than the ancestors of the contemporary English, it cuts the legs out from under a politicized reclamation of the past.[61] In *A Sermon of Reformation*, Fuller's caricature of radical reformers as utopists and quixotic would-be discoverers similarly exemplifies a conservative repudiation of the possibility of positive change, figured by the impossibility of discovery.

In our own era, however, discovery has become the default mode through which knowledge is gained, and one of our underwriting cultural logics. As a foundational assumption of modernity, it has carried with it notions of individual autonomy, the absolute separation of knower from known, the objectification and mastery of that which is discovered, and—as Herbert and perhaps Fuller would have seen it—the separation of knower and known from a Creator who mediates all relationships, including those of knowledge. With modernity, we have become Sauls in relation to knowledge. While Fuller cannot be situated outside the shifts in thought and constructions of boundaries which produced spaces for secularity, he nonetheless registers an awareness of other ways of thinking about the paths to knowledge than those we now take for granted.

---

[61] Schwyzer, *Archaeologies*, pp. 186–97.

Chapter 10
The Discovery of Blackness in the Early-Modern Bed-Trick

Louise Denmead

In the early-modern dramaturgical device of the bed-trick, the "reveal" characteristically hinges on the discovery of a substituted bedmate by the (conventionally) male victim. In this chapter, I will examine what happens to such a discovery when that surrogate sexual partner is a dark-skinned female servant. With primary reference to John Fletcher's *The Knight of Malta* (1617), I will also highlight the significant and, as yet, overlooked connection in the period between the bed-trick motif and the myth of Ixion.[1] The notion of "discovery" in Fletcher's play, and in similar dramatic texts of the period, emphasizes the powerful display of black femininity to the male dupe, and to the audience. At the same time, these particular revelations are characterized not by sudden or automatic recognition, but by the male character's complex and paradoxical *ignoring* of the black female body that can never *quite* be discovered. In his attempts to define black femininity during the process of discovery, the dupe is confounded not only by the black maid's continually shifting persona as enacted on the early-modern stage, but also by a troubling sense of the loss of his own identity.

"What difference twixt this Moor, and her faire Dame?"

In *The Knight of Malta*, Mountferrat, a Knight who has taken a vow of chastity, lusts after the noblewoman Oriana. When she rejects his advances, he tells her dark-skinned Moorish servant, Zanthia, to forge a letter announcing that her mistress plans to marry a Turk—an act considered treasonous. Oriana is then brought to court and defended in a duel between Gomera, an aging Spaniard, and Miranda, a young man in training to become a Knight of Malta. She is found innocent, and Gomera is then given her hand in marriage, though Mountferrat vows to ruin the relationship. Later, when Oriana praises Miranda in the presence of her new husband, Gomera becomes jealous and accuses her of infidelity. Oriana falls unconscious following these allegations. Her husband, wracked with guilt, believes himself to have been

---

[1]   John Fletcher and Francis Beaumont, *Comedies and Tragedies Written by Francis Beaumont and John Fletcher* (London, 1647). This and other plays discussed here will be cited by act, scene and (where possible) line numbers in the body of the text.

the cause of her "death." What has actually happened, however, is that Zanthia (the Moorish servant) has drugged her mistress with a potion that makes the body simulate death. Zanthia reveals her poisoning of Oriana to Mountferrat, and he makes her a promise that once he has raped her mistress as she lies unconscious in her tomb, Zanthia may kill her. The two villains go to the burial-place, but are discovered by Gomera and the others. Finally, Mountferrat is punished by a ruling that banishes both him and Zanthia from Malta, and forces him to marry the Moorish servant. Oriana and Gomera are reunited.

As the above plot summary indicates, Fletcher's play centers on the political stability of Malta. The Turks, described by one Knight as "head-bound Infidels" (I.ii.10), are its threatening outsiders; but the island is also represented as internally divided along racial lines. This schism is particularly evident in the male characters' polarized attitudes to Zanthia and Oriana. As Bindu Malieckal argues: "Zanthia's vicious treatment contrasts with the delicacy and respect with which her female opposite, a white woman named Oriana, is handled. While characters denigrate and attack Zanthia, they praise and rescue Oriana."[2] Though white women's sexual appetite is represented as a threatening and unknowable force requiring patriarchal sanction—after all, without any proof, Oriana is accused of cuckolding Gomera— black serving-women are believed to be doubly voracious, and therefore doubly threatening. In his *Passions of the Minde in Generall* (1604), Thomas Wright draws such a distinction between women in terms of their color and their sexual temperance when he claims that "onely women that be of a hote complexion, and for the most part, those that be blacke or browne ... have their affections most vehement." According to Wright, though all women are subject to "sundry passions," dark-skinned women are subject to even more inflamed passions than their fair English counterparts.[3] When placed in opposition to Zanthia, Oriana is, therefore, persistently purified by her patriarchal culture.

Such portrayals of black female sexuality as especially voracious were, of course, commonplace on the early-modern stage. In John Webster's *The White Devil* (1612), for example, Monticelso condemns Vittoria's "black lust," thus underscoring the commonly perceived link between blackness and carnality (III. i.7).[4] Kim Hall describes an "emerging tradition" in the drama of the early-modern period, in which black maids are seen to act as the evil and promiscuous antitheses of white female virtue.[5] These stereotypical representations of black female sexuality seem to hold true in Fletcher's play. In stark contrast to her mistress, Zanthia openly and bawdily articulates her sexual desires, and Oriana reprimands

---

[2]   Bindu Malieckal, "'Hell's Perfect Character': The Black Woman as the Islamic Other in Fletcher's *The Knight of Malta*," *Essays in Arts and Sciences*, 28 (1999): pp. 53–66; p. 59.

[3]   Thomas Wright, *The Passions of the Minde in Generall* (London, 1604), p. 42.

[4]   John Webster, *Three Plays*, intro. David C. Gunby (London, 1992).

[5]   Kim F. Hall, *Things of Darkness: Economies of Race and Gender in Early Modern England* (Ithaca, 1995), p.188.

her servant several times for this sexual badinage (III.ii.63). Referring to such presumed connections between blackness, transgression and uncontrolled sex drive in the representation of dark-skinned women on the stage, Elliot Tokson notes that black servants often act as a threat to white female chastity by procuring their fair-skinned mistresses in order to facilitate the lust of white men.[6] These conventionally-represented black maids (including Fletcher's Zanthia)—slaves, as it were, to their libidos—often anticipate receiving sexual favors for such procurement. The convention of the lascivious black female servant was, in sum, well-established by the time Fletcher was writing *The Knight of Malta*, and it clearly influences his characterization of Zanthia.

In fact, Gomera excoriates the maid as a "bawd to mischiefe," suggesting his view of her both as a possible panderer and as a sexually transgressive figure. Zanthia fulfills this stereotypical function when she deviously attempts to acquire Oriana's body for Mountferrat's sexual pleasure (IV.ii.266). She boasts to the Frenchman that "it is in me with as much ease / To give her [Oriana] up to thy possession" (IV.i.109). Carolyn Prager says in relation to the generic name Zanthia/ Zanche that "four out of the five bondwomen so named follow the lustful immoral prototype established by Marston in *Sophonisba*."[7] The Zanthia in John Marston's play is Sophonisba's maid, who betrays her mistress in order to facilitate Syphax's lust. I detect a further connection between the two names in the 1647 Fletcher and Beaumont folio, where the stage directions for *The Knight of Malta* confuse the names and refer to Zanthia as "Zanchia" (I.iii.263).[8] It is reasonable to suggest that the name, as well as the character's dark skin color, would have been linked with duplicity and licentiousness for an early-modern audience aware of this theatrical tradition. In fact, while Anthony Gerard Barthelemy shows that black women in early-modern playtexts "seem never to be wholly free of the taint of the powerful and enduring stereotype of the black, malevolent meretrix," he also goes so far as

---

[6]   Elliot H. Tokson, *The Popular Image of the Black Man in English Drama 1550– 1688* (Boston, 1982), p. 84.

[7]   Carolyn Prager, "'If I Be Devil': English Renaissance Responses to the Proverbial and Ecumenical Devil," *Journal of Medieval and Renaissance Studies*, 17/2 (1987): pp. 257–79; p. 276, n. 19.

[8]   She is referred to as "Zanthia" in the first act, but as "Abdella" for the rest of the play. To avoid confusion, I will refer to her as Zanthia throughout. Presumably, the name conflict is due to the fact that the play was a collaborative effort. While authorship is ascribed to both Francis Beaumont and John Fletcher in the 1647 folio, it is more likely written either solely by Fletcher or in collaboration with a playwright other than Beaumont (probably Massinger and/or Ford), as their partnership had ended by 1613. According to Harbage, the play has a performance history dating between 1616 and 1619, when it was acted by the King's Men at the Globe theatre and at Blackfriars. See E.H.C. Oliphant, *The Plays of Beaumont and Fletcher: An Attempt to Determine Their Respective Shares and the Shares of Others* (New Haven, 1927), p. 398; and Alfred Harbage, *Annals of English Drama 975–1700* (London, 1964), pp. 106–7.

to describe Fletcher's Zanthia as "the most malevolent of all the Moorish waiting-women in seventeenth-century drama."[9]

Mountferrat specifically correlates Zanthia's villainy with her dark skin, pointing out that, although his own deeds are nefarious, her blackness predisposes her to more extreme acts of wickedness. He thus locates her blackness in diabolism, while simultaneously naturalizing this association:

> Bloody deeds
> Are grateful offerings, pleasing to the devil,
> And thou, in thy black shape, and blacker actions,
> Being hell's perfect character art delighted
> To do what I, tho' infinitely wicked,
> Tremble to hear. (IV.i.74–9)

Eldred Jones and others have shown how these associations with hell and transgression pervade early-modern dramatic representations of dark skin color.[10] As Tokson, Jones, and Malieckal have noted, the other characters in Fletcher's play consistently associate Zanthia's black skin with the devil, thereby reinforcing the perceived link between blackness, the demonic, and sin.[11] Indeed, Mountferrat denigrates Zanthia as "hell it self confin'd in flesh," confirming the view that Zanthia's black skin is a manifestation of the evil within her (IV.ii.185).

This common perception of black maids as morally reprehensible, and inherently less chaste than their class and racial counterparts, can also be discerned in Mountferrat's polarized view of Zanthia and Oriana. This is based on skin color and social rank, as well as on perceptions of the women's sexual temperance. In a pivotal early scene in the play, Mountferrat alludes to the myth of Ixion in order to reinforce such a disparity between the two women:

> Oh my Zanthia,
> My Pearl, that scorns a staine! I much repent
> All my neglects: let me, Ixion like,
> Embrace my black cloud, since my Juno is
> So wrathfull, and averse; thou art more soft
> And full of dalliance than the fairest flesh,
> And farre more loving. (I.i.180–86)

---

[9]   Anthony Gerard Barthelemy, *Black Face Maligned Race: The Representation of Blacks in English Drama From Shakespeare to Southerne* (Baton Rouge, 1987), pp. 136, 128.

[10]   Eldred Jones, *Othello's Countrymen: The African in English Renaissance Drama, 1550–1688* (Oxford, 1965), p. 48.

[11]   Tokson, *The Popular Image*, p. 61; Jones, *Othello's Countrymen*, p. 82; Malieckal, "'Hell's Perfect Character'," p. 59.

In Greek mythology, Ixion attempted to seduce Hera, but was deceived by Zeus into copulating with a dark cloud disguised in the form of the goddess.[12] Early-modern readers would have been aware of this myth: Sands Penuen fully describes the story of Ixion in *Ambitions Scourge* (1611), and Henry Hutton mentions the myth in his later comic poem *Follie's Anatomie* (1619).[13] Early-modern dramatists also alluded to the myth in their work. In Richard Brome's *The English Moore; or the Mock-Marriage*, Nathaniel sleeps with a woman he thinks to be a negro, but who turns out actually to be a white woman in disguise.[14] His delighted reaction—"I aym'd but at a Clowd and clasp'd a Juno"—clearly indicates that the playwright was drawing on the myth, and it intimates to the audience the character's awareness that a white bed partner is more valuable to him than the black woman who functions as her sexual substitute (V.iii.53).[15] Earlier in the play, when the jealous Quicksands blacks up his young wife Millicent in order to prevent his own cuckoldry, he reassures her by saying "thou dost but case thy Splendour in a Clowd"—again indicating that Brome was invoking the Ixion myth (III.i.88). In this case, Millicent complexly embodies both the dark cloud and the fair woman when she adopts blackface masquerade.

In the scene from Fletcher's play quoted above, while Oriana—Mountferrat's Juno/Hera—is averse to his sexual advances, Zanthia, his "dark cloud," is conversely constructed as more "full of dalliance."[16] A direct link is thereby made between Zanthia's darkness and her sexual impropriety, which in turn is contrasted with the self-restraint associated with Oriana's "fairest flesh." As Kim Hall notes, "blackness" in early-modern discourses is frequently opposed to "fairness," rather than to "whiteness." Such polarized terms, she argues, are most frequently employed in relation to women's physical appearance or to their moral state, and have a particular potency when applied to females.[17] In *The English Moore*, Nathaniel supports this bipartite view of female sexuality when he views the (counterfeit) Moor, Millicent, as *more* easily open to his sexual advances precisely because she

---

[12]   See Apollodorus, *The Library of Greek Mythology*, trans. Robin Hard (Oxford, 1997), p. 142; and Stuart Gordon, *The Encyclopaedia of Myths and Legends* (London, 1993), pp. 85–6.

[13]   Sands Penuen, *Ambitions Scourge* (London, 1611), p. D^r; Henry Hutton, *Follies Anatomie Or Satyres and Satyricall Epigrams with a Compendious History of Ixion's Wheele* (London, 1619), p. E1^v.

[14]   Richard Brome, *The English Moore; or the Mock-Marriage*, ed. Sara Jayne Steen (Columbia, 1983). The dates of the manuscript's composition and first performances cannot be accurately gauged but Steen has estimated that the late 1630s or early 1640s remain the best approximation.

[15]   Though, in a comic turn, when he finds out that the woman in disguise is his spurned lover Phillis, he wishes that she were black again (V.iii.101–2).

[16]   In another allusion to the Ixion myth, Fletcher's Mountferrat conversely curses Zanthia, calling her "thou black swoln pitchy cloud, of all my afflictions" (IV.ii).

[17]   Hall, *Things of Darkness*, p. 9.

is black, thus frustrating Quicksands's attempts to preserve his wife's chastity. On his initial meeting with her, Nathaniel describes Millicent as a "black coneybury" (IV.iii.68), a derogatory term that Sara Jayne Steen identifies as a slang reference for a loose woman, and it clearly indicates that he already assumes, without any assent from Millicent, that she is sexually available to him.[18]

By referring to Zanthia as his "black cloud," Mountferrat is positioning her as Oriana's sexual substitute and as a body that he may sexually enjoy at will. He emphatically makes this point again when he later declares:

> It is not love, but strong Libidinous will
> That triumphs o're me, and to satiat that,
> What difference twixt this Moor, and her faire Dame?
> Night makes their hews alike, their use is so:
> Whose hand so subtile, he can colours name,
> If he do winck, and touch 'em: lust being blind,
> Never in women did distinction find. (I.i.241–7)

In insinuating the interchangeability of women's bodies under the cover of night, Mountferrat's speech can be read as constituting a metatheatrical reference to the popular early-modern dramaturgical convention of the bed-trick. William Bowden defines the trick as follows: "X, expecting to lie with A, is caused to lie with B instead through the conspiracy of A and B."[19] In other words, the dupe has sexual relations with someone other than s/he intended. While there were many variations on the hoax, I will focus here on one particular instance of the bed-trick that involves a noblewoman's attempting to protect her chastity from male lust by putting a black servant in her bed as her replacement.[20]

If we follow the format of the bed-trick device as outlined, Ixion, who is tricked into having sex with a proxy "dark" partner, can be described as a victim of this type of deception. Fletcher's Mountferrat, as a self-styled Ixion figure, could plausibly be interpreted as making this connection with the trick when he muses that Oriana "might with colour dis-allow my suit," which demonstrates his awareness that Oriana may place Zanthia (a woman of color) in her bed to avoid his lust (I.i.18). Though Patricia Phillippy says that "the bed trick relies too much on the willing participation of women, making them partners in crime," we can see that it in fact emphasizes a distinct *lack* of community between women, especially in terms of rank and ethnicity, as the ruse, often engineered by white noblewomen,

---

[18]   Brome *The English Moore*, p. 99, n. 68.

[19]   William R. Bowden, "The Bed Trick, 1603–1642: Its Mechanics and Effects," *Shakespeare Studies*, 5 (1969): pp. 112–23; p. 122.

[20]   There are numerous examples of this particular type of bed-trick in early-modern drama, including John Marston's *Sophonisba* (1605), John Fletcher's *Monsieur Thomas* (c.1610–16), Philip Massinger's *The Parliament of Love* (1624), and Richard Brome's *The Novella* (1632).

is specifically designed to protect culturally-prized white female chastity.[21] Dark-skinned women are therefore doubly othered in this case, by their race and their gender. Black women's sexuality (which, because of their skin color and their inferior rank, is seen to be of inferior value) is clearly sacrificed and exploited in this trick by white men *and* women.

Conventionally, this sort of exchange took place under the cover of darkness, which presumably functioned in this case to conceal the substituted bed-mate's racial identity from the dupe. In addition, night was sometimes specifically figured in early-modern drama as a dark-skinned woman, indicating that discourses of both race and gender are inextricably linked with black women's involvement in the bed-trick. In Thomas Kyd's *The First Part of Jeronimo* (1605), Lazzarato the corrupt courtier anthropomorphically describes night as a black female—"night, / That yawning beldam with her jetty skin"—once more drawing together lasciviousness, blackness, night and femininity (I.i.109–10).[22] In Thomas Dekker's *Lust's Dominion* (1600), the licentious Spanish Queen Mother also personifies night as a black woman who will both permit and disguise her tryst with the Moor Eleazar. Again, this highlights the perception of black females as the loci, and, as mentioned earlier with reference to black maids, the enablers of sexual transgression:

> ... thou spotlesse night,
> Empresse of silence, and the Queen of sleep;
> Who with thy black cheeks pure complexion
> Mak'st lovers eyes enamour'd of thy beauty:
> Thou art like my Moor, therefore will I adore thee,
> For lending me this opportunity,
> Oh with the soft-skinned Negro! (III.i.1–7)[23]

Indeed, Mountferrat describes Zanthia as a "night hag, gotten when the bright Moone suffer'd," which both makes a connection between her dark skin and the darkness of night, and violently asserts that the fact of her birth, which he claims was under the cover of darkness, was itself an aberration (IV.ii.184).

"Women ... memorie ... would one of ye leave me"

Though stoutly asserting Zanthia's and Oriana's interchangeability, Mountferrat in a sense also recognizes that substituting Zanthia for her mistress necessitates a

---

[21]   Patricia Phillippy, *Painting Women: Cosmetics, Canvases and Early Modern Culture* (Baltimore, 2005), p. 94.

[22]   Thomas Kyd, *The First Part of Jeronimo* (London, 1605).

[23]   *The Dramatic Works of Thomas Dekker*, ed. Fredson Bowers, vol. 4 (Cambridge, 1961).

kind of willful ignorance on his part. If he only "winks," he claims, Zanthia's racial difference will become subsumed by darkness. The OED defines "wink" (3) as follows: "to have the eyes closed in sleep, to sleep; sometimes, to doze, to slumber." Garrett Sullivan has examined the relationship between sleep and inattention in the period as well as its association with sensual excess. Mountferrat's bracketing of Zanthia's racial difference necessitates, to use Sullivan's term, a kind of "erotic self-forgetting," when he disregards the racial and social differences between Oriana and Zanthia in order to satisfy his own sexual appetite.[24] Clarindore, in Philip Massinger's *The Parliament of Love*, similarly imagines the disguised and blacked-up Calista as a suitable bed-mate, though, again, only under the cover of darkness:

> The curtains drawn, and envious light shut out,
> The soft touch heightens appetite and takes more
> Than colour, Venus' dressing, in the day-time,
> But never thought on in her midnight revels. (II.iii)[25]

This notion that all women are alike when it comes to sex not only confirms the men's carnality, but also reveals how they essentialize all women's genitalia. As Wendy Doniger argues, "it was precisely women's sexuality that was taken as their essence and that was regarded as essentially the same in all women in the dark, just as their beauty was essentialised and universalized in the light."[26] At the same time, however, the claim that all women are alike is shown to be a tenuous one. Echoing the sentiments of Mountferrat, Clarindore asserts his belief that at night all differences between women, regardless of skin color, may be overlooked; but in so doing he also highlights what Sullivan terms the "foundationality of forgetting" to his sexual relationship with a dark-skinned woman.[27]

William Bowden explains how the conventional bed-trick, dependent as it is on surprise, "is most effective dramatically as a peak, not as a plane … for the more frequently such a meeting was repeated, the more chance there would be for discovery".[28] However, the presence of black maids in these plays complicates the premise of the bed-trick in a way that the presence of a white female would not. Mountferrat, Nathaniel and Clarindore are, in a sense, *willingly* subjecting themselves to deception by deliberately disregarding the social and racial

---

[24]   Garrett A. Sullivan Jr., *Memory and Forgetting in English Renaissance Drama: Shakespeare, Marlowe, Webster* (Cambridge, 2005), p. 41.

[25]   *The Parliament of Love*, in *The Dramatic Works of Massinger and Ford*, ed. Hartley Coleridge (London, 1848). As there are no line numbers in this edition of the text, references are to act and scene only.

[26]   Wendy Doniger, *The Bedtrick: Tales of Sex and Masquerade* (Chicago, 2000), p. 178.

[27]   Sullivan, *Memory*, p. 101.

[28]   Bowden, "The Bed Trick," pp. 120, 119.

differences between black maids and their mistresses. Yet Zanthia's social status and her skin color mean that, unlike their attitude to Oriana's sexual economy, the preservation of Zanthia's chastity is not considered an issue for the men who bed her. When Mountferrat says that Zanthia may be a suitable surrogate to satisfy his lust, he assumes her assent and devalues her chastity in the face of her mistress's, because of both her skin color and her lower rank. As Marliss Desens argues with reference to a bed-trick in *The Decameron*, the replacement of a maid for her mistress "suggests some awareness by the writers of the way in which their society divides women into two groups: the idealized women who are above sex and the despised women who are associated only with sex."[29]

Though willing to have sex with Mountferrat, Zanthia is nonetheless hyper-aware of her disposability and her status as a sexual object: she tells Mountferrat that "like a property, when I have serv'd / Your turnes, You'll cast me off" (I.i.188–9). Mountferrat's desire for sexual pleasure thus marks, not his seduction by Zanthia, but rather his power over black women's sexual economy. In *Black Looks*, bell hooks maintains that desire for a racialized other indicates authority over the other whose body is represented as a locale of temporary sensual pleasure.[30] In just this way, Mountferrat is seen to be content to satiate his lust with black female bodies, which he views as interchangeable with white women's, yet easily discarded. Sexual encounters with Zanthia thus constitute for Mountferrat a temporary deviance from the social order where, under the cover of nightfall, she becomes as desirable a sexual partner as Oriana.

Though Mountferrat is, however temporarily, allowed to disregard the social order through his appropriation of bed-trick tropes, those black-skinned servants coerced into the trick are allowed no such space for forgetfulness. In *Sophonisba*, the eponymous fair-skinned noblewoman attempts to protect herself from the sexual desires of Syphax by placing the drugged and naked black male servant Vangue in her bed. Syphax, "offering to leap into bed, discovers Vangue." Vangue's response to the situation once he comes to consciousness focuses on what he sees as his reversal of fortune:

> Vangue: Where am I? Think. Or is my state advanced?
> O Jove, how pleasant is it but to sleep
> In a king's bed!
> Syphax: Sleep there thy lasting sleep,
> Improvident, base, o'er-thirsty slave.
> *Syphax kills Vangue*
> Die pleased, a king's couch is thy too-proud grave. (III.i.189–95)[31]

---

[29]   Marliss C. Desens, *The Bed-Trick in English Renaissance Drama: Explorations in Gender, Sexuality, and Power* (London, 1994), p. 32.

[30]   bell hooks, *Black Looks: Race and Representation* (Boston, 1992), p. 23.

[31]   John Marston, *The Selected Plays of John Marston*, eds Macdonald P. Jackson and Michael Neill (Cambridge: Cambridge, 1986).

As Virginia Mason Vaughan maintains in her reading of this scene, it is Vangue's fantasy of a higher status (he is "o'er thirsty"), as much as his black skin, that results in his murder.[32] Moreover, Vangue's "forgetting" of his lower social status must be punished by those men who guard their own privileged positions. As both a black man and a servant, Vangue is not allowed any transgression of societal boundaries. The discovery of a black woman in the bed-trick, however, produces a far more ambivalent reaction, a point to which I will now turn.

In his discussion of *All's Well That Ends Well*, John Wain argues that "love feeds on recognition and knowledge of the loved person; lust, by contrast, is blind; its patterns are laid down in advance and it feeds on whatever approximates to those patterns."[33] Indeed, Mountferrat himself at one point claims that "I am … almost blind, and deafe. / Lust neither sees nor hears ought but it selfe" (I.i.169–70). The lascivious Frenchman categorically states that he does not love Zanthia; for him she functions as a sexual plaything that can be cast off at his whim. In a similar turn to Mountferrat, Nathaniel in *The English Moor* admits that by having sex with a black woman he will prove that he has "as good stomach[s]" as her partner Quicksands, denoting not only his sexual appetite for her, but also his culture's assumption that not every white man can "stomach" having sex with a black woman (IV.iii.106). Dark-skinned females are paradoxically viewed as both sexually desirable and repugnant. Fletcher's *Monsieur Thomas* (c.1610–16) addresses this contradictory reaction more fully. Here, Thomas is tricked into getting into bed with Kate, a black maid. However, candlelight reveals the woman's true identity, and he cries out:

> Holy Saints defend me!
> The devil, devil, devil! O the devil! …
> I am abused most damnedly, most beastly;
> Yet if it be a she-devil … (V.iv.50–51, 53–4)[34]

The stage direction for the *Monsieur Thomas* bed-trick reads "a bed discovered with a Black More in it." When Thomas "discovers" a black maid in place of the woman he intended to seduce, the dupe immediately names her "devil," thus attempting to intercede between her and the audience: to define and control who, or what, she is (or is not). However, his discovery ends not necessarily in his complete appropriation of the servant's otherness, but rather in Thomas's bracketing, perhaps even his deliberate *forgetting*, of his initial interpellation of her. His discovery of Kate produces a tension between his desire to sexually possess her and his acute fear of her racial difference. This culminates in the fragmentation of Thomas's own

---

[32]   Virginia Mason Vaughan, *Representing Blackness on English Stages, 1500–1800* (Cambridge, 2005), pp. 87–8.

[33]   John Wain, *The Living World of Shakespeare: A Playgoer's Guide* (Louisiana, 1966), pp. 120–21.

[34]   John Fletcher, *Monsieur Thomas* (London, 1639).

identity, wavering as it does between desire and anxiety. Such ambivalent reactions to dark-skinned females can be seen to pervade wider discourses of "discovery" in Renaissance travelogues, and, as historian Jenifer L. Morgan has shown, European male travel writers routinely enact this very "ideological maneuver" when they juxtapose the beauty and the monstrosity of the foreign black female body.[35]

The ambivalence at the heart of Thomas's discovery is further amplified by early-modern theatrical modes of representation, which make Kate—a male actor in drag and blackface—an exceedingly contrived and evasive persona. Though aware of the presence of Kate prior to Thomas's own discovery of her, the audience is also implicated in "forgetting" the black maid's "real" identity. The image of the doubly-othered black woman, mentioned earlier, can be seen as particularly pertinent here to considerations of early-modern theatrical representations of black femininity, which constitute a double absence and highlight what Lynda Boose terms the "unrepresentable"—and, perhaps, for Thomas and the audience, "undiscoverable"— nature of black women.[36]

"Go leap her, and engender young devillings"

In a period when black women, to use Jennifer Morgan's terms, "mark metaphorically the symbiotic boundaries of European national identities and white supremacy," the presence of the black female body in the bedroom would have represented the ultimate taboo.[37] As Virginia Mason Vaughan astutely notes, there are no consummated bed-tricks featuring black women in the period: "while the spectre of miscegenation is raised, it is never literalised on the stage."[38] The risk to social hierarchies thought to be posed by sex with a dark double in the bed-trick is also a concern in early-modern texts that reference the Ixion myth. In Eldred Revett's poem "One Enamour'd on a Black-Moor" (1657), for example, the black woman desired by the male speaker is imagined as "the goddesse's deputed cloud" and any contact between them, he predicts, will result in "a dapple race."[39] In Penuen's version of the myth in *Ambitions Scourge*, Ixion is said to have displayed presumption in attempting to seduce Hera. The false Hera with whom he actually lay, moreover, bore Ixion the outcast child Centaurus. The latter,

---

[35]   See Jennifer L. Morgan, "Some Could Suckle Over Their Shoulder: Male Travelers, Female Bodies, and the Gendering of Racial Ideology, 1500–1770," *The William and Mary Quarterly*, 3rd ser., 54/1 (1997): pp. 167–92.

[36]   Lynda E. Boose, "'The Getting of a Lawful Race': Racial Discourse in Early Modern England and the Unrepresentable Black Woman," in Margo Hendricks and Patricia Parker (eds), *Women, "Race" and Writing in the Early Modern Period* (London, 1994), pp. 35–54; p. 47.

[37]   Morgan, "Some Could Suckle, p. 169.

[38]   Vaughan, *Representing Blackness*, p. 91.

[39]   See Hall, *Things of Darkness*, p. 288–9, ll. 42, 44.

grown to manhood, is said to have produced a hybrid race when he sired horse-centaurs on Magnesian mares:

> But yet Ixyon on this seeming Faire,
> (Which was indeed nought but delusive aire)
> Begot the Centaures, that rebeld gainst Jove:
> The mixture monstrous, so th'effect did prove:
> But of Ixyon's issue, write who will,
> Monsters in minde (not Nature) move my quill. (D$_2$^{v})

Penuen's allusion to the commixing of the black cloud and Ixion as "monstrous" imagines their offspring as a hybrid race that he views as distinct from, and inferior to, creatures of the natural world. Kathryn M. Brammall has outlined how, though monsters continued to be associated with physical deformity throughout the early-modern period, the rhetoric of monstrosity in the sixteenth and seventeenth centuries became increasingly focused on deviant behavior.[40] As Judith J. Kollermann notes, the figure of the centaur is explicitly associated with rapacity, "the result of presumptuous and unlawful appetite, [and] is himself the inheritor of perverted sexual appetite, and the final result is a race of beings that manifest impulsive violent behaviour and unlawful desires."[41] The act of interracial sex between Zanthia and Mountferrat is also believed to be as monstrous as the hybrid creatures it is anticipated to produce. Norrandine tells Mountferrat: "Away French Stallion, now you have a Barbary Mare / of your own, go leap her, and engender young devillings" (V.ii.314–15).[42] His image of Zanthia as "a Barbary Mare" clearly recalls Iago's warning to Brabantio when his daughter clandestinely marries the Moorish general Othello: "you'll have your daughter covered with a Barbary horse; you'll have your nephews neigh to you; you'll have coursers for cousins and jennets for germans" (I.i.109–12).[43]

The particular reference to horses in the case of Norrandine's invective, mentioned above, points not only to what he sees as the bestial nature of the relationship between Mountferrat and Zanthia, but also to the threat to patriarchal social order that is specific to the involvement of black maids in the bed-trick. As

---

[40]   Kathryn M. Brammall, "Monstrous Metamorphosis: Nature, Morality and the Rhetoric of Monstrosity in Tudor England," *The Sixteenth Century Journal*, 27/1 (1996): pp. 3–21.

[41]   Judith J. Kollermann, "The Centaur", in Malcolm South (ed.), *Mythical and Fabulous Creatures: A Source Book and Research Guide* (London, 1987), p. 230.

[42]   The *OED* gives one definition of stallion as "a man of lascivious life" (2b). It is also used in this sense in *The English Moore*, where Quicksands refers to "the Lustfull Stallions of our time" (III.i.48).

[43]   Shakespeare, *Othello*, ed. Julie Hankey, 2nd edn (Cambridge, 2005).

Wendy Doniger argues, stallions are said to "service," but also to "serve," mares.[44] Norrandine's reference to Mountferrat as a "stallion" and Zanthia as a "mare" may thus also anxiously point to the challenge that Zanthia's color is believed to pose to patriarchal notions of inheritance. Lynda Boose shows how, in a society that depends on a principle of inheritance in which the father's identity—whether that be his name, his authority, or his property—is passed on to his son, the idea that a black woman may possess the authority to influence the color of her child provokes extreme anxiety.[45] Indeed, Boose notes a curious silence in travelogues where miscegenous relationships featuring black women are concerned. In the period, the ability of the black matrix to attenuate a fair child's skin color, she argues, emphasizes not only the power of blackness but also a threatening reversal of gender hierarchies, such that the child resembles the mother, and not the father.[46] Interracial sex between a black woman and a white man was thus viewed as something that destabilized both gender and racial hierarchies. When the woman involved in interracial sex is both dark-skinned and of a lower rank than her lover, as is evident in the case of the black maid's involvement in the bed-trick, her believed ability to determine the color of her offspring from the encounter is viewed as triply threatening: it may destabilize hierarchies of gender, race *and* rank.

When confronted with the body of a black woman in the place of her mistress, male bed-trick dupes experience a complex and conflicted mix of sexual desire and loathing. Though he views Zanthia as a sexual object that he may easily dispose of, Mountferrat nonetheless sees himself as essentially altered by his relationship with this black servant: "thou hast made me more devil than thy self," he claims (II.iii.20–21). Here, Mountferrat articulates a fear of his and Zanthia's *sameness* as much as his anxiety over the threat of Zanthia's racial alterity. Like Thomas, Mountferrat experiences the fragmenting of his identity through his contact with a Moorish servant. However, it is in the dénouement of the play that Zanthia's threat to the social order is most profound. Valetta (a Knight of Malta) tells the Frenchman: "your doom is then / To marry this coagent of your mischiefes" (V.ii.309–10). *The Parliament of Love* has a similar ending: Clarindore is ordered to marry Calista, a Moorish servant, as punishment for his sexual misconduct, a

---

[44]   Doniger, *The Bedtrick*, p. 179. The *OED* defines "serve": "To cover (the female) esp. of stallions, bulls etc. kept and hired out for the purpose" (def. 52). Barnabe Googe's translation of Conrad Heresbach's *Husbandry* (1577) is cited as having used the word in this sense.

[45]   Boose, "'The Getting'," pp. 45–6.

[46]   *Ibid.* Celia Daileader similarly notes, with reference to cross-racial relationships featuring black women, that "they all work according to the same curious alchemy: white + black = black." See Daileader, *Racism, Misogyny, and the Othello Myth: Inter-racial Couples from Shakespeare to Spike Lee* (Cambridge, 2005), p. 16. See also Harryette Mullin, "Optic White: Blackness and the Production of Whiteness," *Diacritics*, 24/2–3 (1994): pp. 71–89; p. 73.

penalty that he views as "cruelty / beyond expression." While he is willing to have sex with her, the legitimization of the relationship through marriage is abhorrent to him (V.i.513). To Clarindore's relief, however, "Calista" turns out to be his spurned wife Beaupre, masquerading as a Moor. At the end of *The English Moor*, Nathaniel also discovers that the black woman he has slept with is actually his spurned lover in blackface disguise. In a sense then, while the villainous male characters may temporarily "forget" the difference between white women and their black counterparts, they, and indeed the audience, ultimately recall that these types of relationships are anathema to social norms. What makes *The Knight of Malta* so remarkable, then, is that the forgetting of difference is violently enforced, as a social punishment, at the end of the play. In an ironic reversal of fortune, Mountferrat finds himself to have become interchangeable with Zanthia (his "black cloud"), and the victim of his own ruse. Norrandine curses both of them in similarly demonic terms: "We'll call him Cacodemon, with his black gib, / There, his succuba, his devils seed" (V.ii.209–10).

Conclusion

Producing an object of sexual desire that is also viewed as a repugnant devil, the discovery of a dark-skinned maid in the bed-trick necessitates a complex and paradoxical mental act on the part of the male dupe and the audience. The ambivalence that lies at the centre of the male dupe's discovery is one that is specific to the black woman's involvement in this dramaturgical device, and it is exemplified by theatrical modes of representing black women on the early-modern stage. At the same time as the dupe discovers her, or attempts to name, or define, what she is, the doubly-absent black female figure continually disappears and constantly evades full representation. Yet the threat she poses to patriarchal notions of signification and inheritance, indeed to male identity itself, remains. Though Zanthia is physically absent at the end of the play, as the two villains are banished from Malta, the power of her blackness remains a vestigial threat that is always *ideologically* present; always both discovered, and undiscoverable.

Chapter 11
Newness and Discovery in Early-Modern France

Vincent Masse

A telling example of the sixteenth-century adaptability of the term "New World," aptly illustrating how the concepts of "discovery" and "newness" can be challenged, appropriated, and played with, appears in Charles Fontaine's *Les nouvelles, & Antiques merveilles* (1554). Fontaine argues that the New World was not discovered by Christopher Columbus, or by Amerigo Vespucci, but by a Frenchman named "Béthencourt":

> *ie croy, & tiens que Betencourt y fut le premier, non pour favoriser à ma nation (car Betencourt estoit Françoys) ains par ce que le [sic] date que ie trouve aux livres latins, quand Americ trouva lesdictes terres, est depuis, & long temps apres la date que ie trouve que Betencourt y fut.*

> [I believe, and hold that Béthencourt was the first there, not in order to favor my own nation (Béthencourt was a Frenchman), but because the year I find in Latin books, on which Amerigo found said islands, is later, and very well after the year on which I [also] find that Béthencourt went there.][1]

The dispute rests on what lands or islands qualify as being part of the "new" world, for Jean de Béthencourt indeed "discovered" and partly conquered the Canary Islands, well before Columbus or Vespucci ever set foot in America. François de Belleforest, also seeking to establish the pre-eminence of French explorers, uses a strategy antithetical to that of Fontaine. In *L'Histoire universelle* (1570), instead of conflating the Canary Islands with the Americas, Belleforest dissolves the new "Occident" into its constituents. While conceding the discovery of "Mexique," "Peru," and "Cusco" to the Spaniards, he makes sure to attribute that of "Floride," "Canada," and "Labradour," to Jacques Cartier, and one anonymous nobleman from Milleraye.[2]

"Discovery" is a conceptual argument related to authority, quite effective in legitimizing a claim, and all the more so in a period of nascent copyrights and

---

[1]   Charles Fontaine, *Les nouvelles, & Antiques merveilles* (Paris, 1554), f. 3r°. Unless otherwise noted, translations are my own throughout this chapter.

[2]   François de Belleforest, *L'Histoire universelle du monde* (Paris, 1570), p. 247.

legally approved conquests. Used bluntly or subtly, it can undermine foreign or past achievements—sometimes under the guise of praises—and promote or justify one's own relevance next to the Portuguese and Spanish empires, the Ancients' knowledge, or the medieval encyclopedic monuments. What I will consider in the next few pages, however, is not how "newness" as a discursive artifact is granted or refused to variously contested acts of discovery and/or invention; but rather how "newness" and "discovery," and the early-modern invention thereof, intersect with the advent of print culture in the fifteenth and sixteenth centuries and beyond. "Newness," as a rhetorical argument, helps push forward new discourses—or old texts under new guises—and new methods of distribution; it is a license to print, and an incentive to read.

Incremental Newness and Print Culture: Addenda, Serializations, Periodicals

In the mid-sixteenth century, the French literary world came into the grip of a best-selling craze, imported from Spain, and yet promptly "Frenchified":[3] the ever-expanding series of the *Amadis de Gaule*. The Spanish original printed series, launched by Garci Rodríguez de Montalvo, ran from books I (1508) to XII (1546). Spin-offs and additional sequels were numerous and quick to follow. The first eight books of the French remake were penned by Nicolas Herberay des Essarts, and came out at roughly one-year intervals between 1540 and 1548. Books IX–XIV, "translated" by an array of authors, were published in 1551–74. Books XV–XXI followed in 1577–81, and XXII–XXIV in 1615, respectively adapted from Italian and German sequels.

As Virginia Krause has recently argued, the French *Amadis* marks the advent of a new romance format, quick to be imitated and criticized: *serialization*.[4] Starting with the fourth installment, "narrative modes begin to encourage readers to keep reading from one book to the next, using each new book to program a desire for a sequel."[5] The fifth book's prefatory verses could hardly be more explicit in this regard:

> *Quand d'Amadis j'ay veu le Premier livre,*
> *Il me fait estre amoureux du Second,*
> *Et ceste amour ne me veult laisser vivre*
> *Sans voir le Tiers, tant me semble facond.*
> *Et puis ce Tiers, qui au Quart me semond,*

---

[3]   Marian Rothstein, *Reading in the Renaissance:* Amadis de Gaule *and the Lessons of Memory* (Newark, 1999), p. 51 and *passim*.

[4]   Virginia Krause, "Serializing the French *Amadis* in the 1540s," in M. Rothstein (ed.), *Charting Change in France around 1540* (Selinsgrove, 2006), pp. 40–62.

[5]   *Ibid.*, p. 45.

Me fait plus fort desirer le Cinquiesme.
[When I saw Amadis's first book,
It made me fall in love with the second:
And this love does not want to let me live,
Without seeing the third, so fruitful does it seem:
And then this third, which summons me to the fourth,
Makes me desire the fifth even more strongly.][6]

Serialization and desire for hitherto unpublished material entails a posit of what we might call *incremental newness*; not only as a positive attribute, but also as a criteria of pertinence to publish and of desire to read. As Michel Simonin observes, the *Amadis* series "asserts itself as *novel*, or better yet as provoking the steady updating of *newness*" through "the *extensions, sequels,* or simply *volumes,* supposedly engendered as if by the genealogical proliferation of the characters."[7]

Notwithstanding that *Amadis* may indeed be the first French literary narrative to "consciously and constantly be promoted by virtue of its newness,"[8] the occurrence of texts and paratexts positing newness as a meliorative attribute is well distributed among period genres and formats.[9] Early print editions of medieval and contemporary chronicles, for example, are highly susceptible to arguments relying on the desirability of updating processes. Robert Gaguin's *Compendium de origine et gestis Francorum* (1495) kept on being published, revised, and expanded, well beyond the death of its author in 1501. Nicole de la Chesnaye's French translation, titled *Les grandes chroniques,* appeared in 1514;[10] its first and subsequent editions each are printed with new additions covering events up to the issuing year, as extolled on title pages. Most of the new material is penned by Pierre Desrey, an experienced *continuateur* who also authored the 1513 supplements to the French translation of Werner Rolevinck's *Fasciculus temporum.*[11] Desrey's additions appear at the very end of every edition consulted, in the format of chronologically stratified appendices.

---

[6]   *Le Cinquiesme Livre d'Amadis* (Paris, 1550), f. 3v°; translation from Krause, "Serializing," p. 47.

[7]   Michel Simonin, "La disgrâce d'*Amadis,*" in *L'Encre et la lumière* (Geneva, 2004), p. 234. See also Howard Bloch, *Etymologies and Genealogies: A Literary Anthropology of the French Middle Ages* (Chicago, 1986), pp. 93–4.

[8]   Simonin, "La disgrâce d'*Amadis,*" p. 234.

[9]   "Paratext" refers to any textual or pictorial element on the fringe of the text itself: title page, preface, indexes, etc., as defined by Gérard Genette in *Seuils* (Paris, 1987).

[10]   See Franck Collard, "Histoire de France en latin et histoire de France en langue vulgaire: la traduction du *Compendium de origine et gestis francorum* de Robert Gaguin au début du xviᵉ siècle," in Y.-M. Bercé and Ph. Contamine (eds), *Histoires de France, historiens de la France* (Paris, 1994), pp. 91–118.

[11]   *Ibid.,* p. 97.

From the 1518 edition onwards, Gaguin's chronicles are retitled *La mer des croniques et mirouer hystorial de France*, possibly in order to gain market share over another lucrative yet aging collection of chronicles and addenda: *La mer des histoires*, which offers a further example of the display of incremental newness. "*Mer des histoires* [sea of stories]" being a locution quite common in late fifteenth- and early sixteenth-century titles, it is difficult to establish how many versions exist, and which ones are directly related to Jean de Columna or Giovanni Colonna's *Mare historiarum*, or to the *Rudimentum novitiorum* first printed at Lübeck in 1475. Greater confusion yet abounds concerning the various continuations of *La Mer*, as if the anonymity of the source material particularly allowed the multiplication of alternate versions, additional volumes, and revisions. And yet—as with Gaguin's chronicles—one element is systemic: the professed commitment to continual updates, up to the current year. The 1536 and 1543 editions, respectively published by Galliot du Pré and C. Langelier, pledge to offer "*toutes les hystoires, actes et faictz dignes de mémoire puis la création du monde jusques en l'an mil cinq cens. xxxvi [all the stories, events, and deeds worth remembering, from the Creation of the World up to the year 1536]*," and "*jusques en l'an mil cinq cens xliii [up to the year 1543]*." Editors also unilaterally added new volumes to the original two. An anonymous Parisian printer thus offered *Le quatriesme livre de la mer des hystoires [The fourth volume of the Sea of Stories]* as early as 1518; yet in 1550 Jean Le Gendre's newly composed addition is titled *Tier livre de la fleur et mer des hystoires [Third volume of the Flower and Sea of Stories]*.

Additions to Gaguin's chronicles and to *La mer des histoires* borrow both structurally and literally from Eusebius of Caesarea's *Chronicon*. Eusebius died c.339 CE, yet, similarly to Gaguin's *Compendium*, the updating of his *Chronographia*, as preserved in Jerome's Latin translation, far survived his demise. Jerome himself filled in the years 326–78; Sulpicius Severus expanded it to 403; Paulus Orosius, to 417; Prosper of Aquitaine, to 455; Idiatus, to 468; Marcellinus, courtier of the Emperor Justinian, to 534; Victor Tunnunensis or Tonnennensis, Bishop of Tunis, to 566; John of Bisclaro or Bisclarum, to 590; etc.[12] Emerging in the 1470s, printed editions of Eusebius's *Chronicon* continued the tradition of offering new material. Henri Estienne's 1512 edition thus includes Matthias Palermus's update to 1481, while the period 1482–1512 is covered by Johannes Multivallis of Tournai; both additions are prominently announced on its title page.[13] Moreover, the *Chronicon* provided a typographical framework—numbers and columns[14]—which not only was followed by subsequent works, such as Achilles Pirminus Gasser's *Historiarum chronicorum mundi epitome* (1532), or Jean du Tillet's *La chronique des roys des France ... jusques en l'an 1551* (1551); but

---

[12]   See Bloch, *Etymologies and Genealogies*, p. 38.

[13]   Eusebius, *Eusebii Caesariesis Episcopi Chronicon* (Paris, 1512).

[14]   See Anthony Grafton, *What Was History? The Art of History in Early Modern Europe* (Cambridge, 2007), p. 175.

also provided a remarkably undemanding technique of integrating the "discovery" of America into both pre-existing and newly composed chronologies. Gasser's *Epitome*, updated and translated into French, thus gives, next to the heading "1492," this straightforward notice, conveniently free of details: "*Aucunes Isles en la mer Oceane totallement incongneues des anciens / furent en ce temps trouvees comme ung nouveau monde*" [In those times were found, as a New World, some Islands in the Oceanic sea unknown to the Ancients].[15] As Elizabeth Eisenstein asserted, Europe's printing revolution featured throngs of master texts, widely distributed and yet easily amended: new discoveries could be incorporated into newer editions, and thus printing contributed to the invention of an accumulative research tradition.[16]

Eisenstein's argument is a reply to that of Lucien Febvre and Henri-Jean Martin, who observed that the advent of printing at first massively featured medieval texts, as the needs of the new presses could only be met by the old manuscripts.[17] Overlooking "the possibility that an increased output of old texts may have contributed to the shaping of new ideas,"[18] Febvre and Martin consequently postulated that the early-modern printing industry bestowed cultural inertia, through the accelerated diffusion of "ancient bias" and "appealing falsehoods," notably as pertaining to geographical data.[19] Marshall McLuhan, however, contended that even if "the first two centuries of print culture were almost wholly medieval in *content*," it was not a new message, but a new medium that induced major epistemic shifts in early-modern Europe.[20] That newness-related arguments are on display even in the paratext of aging, yet newly printed, editions of medieval texts—whether or not the latter are updated—corroborates McLuhan's argument: the staging of "newness," and of its "discovery," are ostensibly functions of print-culture processes themselves.

With the advent of printing, even age-old medieval texts such as pseudo-Aristotle's *Secretum secretorum* or the anonymous *Secretz de la nature* were capable, through vernacularization and greater diffusion, of helping to implement a "shift in consciousness" signaled by "the publication in the sixteenth century of scores of 'books of secrets' that professed to reveal, to anyone who could read them, the secrets of nature and the arts."[21] Older texts were pushed forward

---

[15] Achille Pirminius Gasser, *Brief recueil de toutes Chroniques & Hystoires* (Anvers, 1534), unpaginated, under "1492."

[16] Elizabeth Eisenstein, *The Printing Press as an Agent of Change* (Cambridge, 1980).

[17] Lucien Febvre and Henri-Jean Martin, *L'Apparition du livre* (Paris, 1958).

[18] Eisenstein, *Printing Press*, p. 85n.

[19] Febvre and Martin, *L'Apparition*, pp. 386, 388–90.

[20] McLuhan, *The Gutenberg Galaxy: The Making of Typographic Man* (Toronto, 1962), p. 184.

[21] William Eamon, "From the Secrets of Nature to Public Knowledge," in D.C. Lindberg and R.S. Westman (eds), *Reappraisals of the Scientific Revolution* (Cambridge,

under new guises, as though newly exposed, and in some instances the strategy appears to have been malicious: Marco Polo's *Description geographique ... nouvellement reduict en vulgaire François* [*Geographical description ... newly rendered in the French vernacular*] (Paris, 1556) is falsely portrayed as the very first French version of Polo's travels—an astonishing claim; while Gossuin de Metz's *Mirouer du monde nouvellement imprime* [*Newly printed mirror of the World*] (Geneva, 1517) is represented as having been newly composed by one "Francoys Buffereau" over the years 1514–16, even though Gossuin's thirteenth-century *Image du monde* had already been printed in Paris in 1501.

It should therefore not be overly unexpected that the "genealogy" of texts supplementing Eusebius's *Chronicon*—which, according to Howard Bloch, participates "in a medieval epistemology of origins by which truth and value" are grounded "at their source, and in the very idea of source"[22]—would nonetheless supply early-modern Europe with a first prototype for later newspapers, in which every "supplement" was eventually set to be as independent and as "fresh" as possible. The counterintuitive metamorphosis of a text relying on its ancestry into texts relying on newness is a further example of old texts producing new effects. Specific stages in this metamorphosis can be identified. Thus, while his aim is completeness rather than contemporaneousness alone, Pierre Desrey, relieved through Gaguin's "preliminary work" of having to write his way up to 1500, nonetheless qualifies as a practitioner of *"histoire immédiate"* [history as it unfolds].[23] His additions are, however, sold as bound to Gaguin's *Compendium*. This is a constraint that the later, anonymous *Quatriesme livre de la mer des hystoires* (1518) is relieved of, since the "installment" is published separately. And so it is with every book of the *Amadis de Gaule* series.

While the advent of stand-alone, periodical issues *per se* is an invention, throughout Europe, of the early seventeenth century,[24] the immediate antecedent to Théophraste Renaudot's *Gazette*—which started in 1631 and is usually

---

1990), p. 340. See also John Block Friedman, *"Secretz de la Nature [Merveilles du Monde]*, Les," in J.B. Friedman and K. Mossler Figg (eds), *Trade, Travel, and Exploration in the Middle Ages: An Encyclopedia* (New York, 2000), pp. 545–6.

[22]   Bloch, "Genealogy as a Medieval Mental Structure and Textual Form', in H.U. Gumbrecht et al. (eds), *Grundriss der Romanischen Literaturen des Mittelalters*, 11/1 (Heidelberg, 1986), p. 136.

[23]   See Jean Lacouture, "L'histoire immédiate," in J. Le Goff and R. Chartier (eds), *La nouvelle histoire* (Paris, 1978), pp. 270–93; and Claude Thiry, *L'Histoire immédiate: une invention du Moyen Age?* (Liège, 1984), and "Historiographie et actualité (XIVᵉ et XVᵉ siècles)," in H.U. Gumbrecht et al. (eds), *La littérature historiographique des origines à 1500*, 1/3 (Heidelberg, 1987), pp. 1025–63.

[24]   See Brendan Dooley, "Introduction," in B. Dooley and S.A. Baron (eds), *The Politics of Information in Early Modern Europe* (London, 2001), pp. 8–9.

considered to be the first French (weekly) newspaper[25]—is yet another series of addenda to chronicles. Pierre Victor Cayer, also known as Palma Cayer, published his *Chronologie septénaire*, covering the years 1598–1604, in Paris, 1605. Its "prequel," the *Chronologie novenaire*, followed three years later, covering the years 1589–97. Jean Richer, Cayer's editor, later supplied a sequel periodical to the *Chronologies*, titled *Le Mercure François* [*The French Mercury*], or bearer of news. Twenty-five volumes were published between 1611 and 1648.[26] From the fifteenth to the seventeenth centuries, the warrant for supplemental material to chronicles thus shifted from a standard of "completeness" to one of "newness." Rather than referring directly to specific content, "newness" was a discursive, medium-related argument of pertinence.

Rituals of Discovery

Discursive "unprecedentedness" is staged in accordance with specific constraints pertaining to genres and formats. Those constraints, as well as texts and paratexts arguing "newness" as a meliorative attribute, are particularly remarkable when the argument is demonstrably false. On the recycling of material in sixteenth-century anthologies of *nouvelles* [short stories]—a genre in which topicality and quirkiness are especially cherished—Gabriel-A. Pérouse observed that editors commended the freshness of the featured stories even when they were borrowed, and already well-known. On account that "*nouvelles* are substantially and necessarily associated with topicality," assertions of newness "are considered to be essential by the editors" and "loudly proclaimed"; "it [thus] makes more sense to lie" about actual newness "rather than risk falling into" the nonsensical and non-existing genre that "non-topical *nouvelles*" would have been.[27] In chronicles and related addenda, it is temporal and geographical comprehensiveness that asserts "newness" and relevance, as illustrated by Jean Longin's paratextual strategy in his edition of Jean Le Gendre's aforementioned *Tier livre*—which Longin issued simultaneously to the first two (updated) volumes of the *Mer des histoires*. Longin deploys both temporal and geographical comprehensiveness as a pertinence (or marketing) argument. Even though his version of the first two volumes of *La mer des histoires* covers events "*iusques en l'an Mil cinq cens. l.* [*up to the*

---

25   Note however that Jean Martin and Louis Vendosme's *Nouvelles ordinaires de divers endroicts*—the very early issues of which may have been manuscript—started almost simultaneously to Renaudot's *Gazette*. See Louis Trenard's chapters in C. Bellanger, J. Godechot, P. Guiral, and F. Terrou (eds), *Histoire générale de la presse française* (Paris, 1969).

26   Trenard, *Histoire générale*, pp. 78–9.

27   Gabriel-A. Pérouse, *Nouvelles françaises du XVIe siècle* (Geneva, 1977), pp. 115–16. See also Deborah N. Losse, *Sampling the Book: Renaissance Prologues and the French Conteurs* (Lewisburg, 1994), p. 58.

year 1550]," the 1543–50 additions are said to privilege "*les choses faictes &*
advenues en France [*events which occurred in France*]." Le Gendre's *Tier livre*
supposedly and conveniently offers the outstanding international news: "*la fleur et*
mer des hystoires plus celebres & memorables advenues tant en l'Asie & Affricque
qu'en l'Europpe ... commençant en l'an mil cinq cens trente & cinq, & continuant
iusques en l'an mil cinq cens cinquante & ung [*The Flower and Sea of Stories*
most famous & worthy of memory, that happened in Asia & Africa, as much as
in Europe ... starting from the year 1535, and continuing up to the year 1551]."[28]
However, once past the title page, Longin's strategy comes undone: Le Gendre
does not conceal his preference for French and—to a smaller extent—European
events.[29] Moreover, and notwithstanding its title, Le Gendre's chronicle ends in
1550. The importance of topicality and comprehensiveness, in the genre of the
chronicles, is aptly demonstrated by the desirability of inaccurate assertions to
that effect.

However, the format employed to convey newness and to emulate the act
of discovery is first and foremost that of the *occasionnel*: short, cheap relations
of recent and important events, such as battles, coronations, funerals, and royal
entries.[30] The *occasionnel* format, incidentally, is another canonical antecedent
to seventeenth-century periodicals:[31] Jean-Pierre Séguin, pioneer in the study of
canards and *occasionnels*, alternatively called them "non-periodical information
bulletins."[32] Printed news bulletins, or topical literary pieces, are as old as printing
itself, and facilitated the advent of the new medium,[33] much as variously serialized
and renewable texts, such as calendars, almanacs, and ephemerides, aided the
fifteenth-century onset of print culture. Throughout Europe, their production
outputs cluster around specific events: the fall of Negroponte to the Turks (1470),[34]

---

[28]   *Le premier volume de la mer des histoires ... & le second ensuyvant* (Paris, 1550),
Jean Le Gendre, *Tier livre de la fleur et mer des hystoires* (Paris, 1550).

[29]   Gaguin's *Compendium*, first written as a history of France, will similarly later be
marketed as covering "Universal History"; see Collard, "Histoire de France," p. 111.

[30]   See Pascal Lardellier, *Les miroirs du paon: rites et rhétoriques politiques dans la*
France de l'Ancien Régime (Paris, 2003).

[31]   Trenard, *Histoire générale*, pp. 29–41; Roger Chartier, "Le périodique: les
antécédents," in H.-J. Martin and R. Chartier (eds), *Histoire de l'édition française*, vol. 1
(Paris, 1982), pp. 411–12.

[32]   Jean-Pierre Séguin, "Les feuilles d'information non périodiques, ou 'canards', en
France," *Revue de synthèse*, 3rd ser., 78/7 (1957): pp. 391–420.

[33]   Pierre Civil and Danielle Boillet (eds), *L'Actualité et sa mise en écriture aux XVᵉ–*
XVIᵉ et XVIIᵉ siècles: Espagne, Italie, France et Portugal (Paris, 2005), p. 8. Note that
hand-written newsletters, such as the Italian *avvisi* or the German *zeitungen*, long existed
independently to the "printing revolution"; see Françoise Weil, "Les gazettes manuscrites
avant 1750," in *Le Journalisme d'Ancien Régime* (Lyon, 1982), pp. 93–100.

[34]   Margaret Meserve, "News from Negroponte: Politics, Popular Opinion, and
Information Exchange in the First Decade of the Italian Press," *Renaissance Quarterly*,
59/2 (2006): pp. 440–80.

the War of the League of Cambrai (1508–16),[35] the Sack of Rome (1527),[36] Charles V's conquest of Tunis (1535),[37] etc. *Occasionnels* are not unrelated to chronicles or large ensembles in that many exotic news bulletins, such as Niccolò de' Conti's 1414–39 travels to South and South East Asia, and *L'Histoire du Nouveau Monde descouvert par les Portugaloys*—a short, 12-sheet octavo published in 1556—are in fact off-prints taken out of heavy volumes, respectively Poggio Bracciolini's *De varietate fortunae* and Pietro Bembo's *Historiae Venetae* (1551). The opposite also holds true: past its immediate newsworthiness, Columbus's 1493 *Epistola* found its way into many chronicles and anthologies, throughout the sixteenth century and beyond.

The most famous fifteenth-century *occasionnel* is undoubtedly Columbus's "Letter to Santangel," which survives in 12 editions from 1493 to 1497, in a format ranging from three to seven leaves.[38] Yet the staging of the act of discovery is in fact quite common in news bulletins, from the sighting and report of comets, deformed stillborns, and fallen cities, to exclusive prophecies and the promulgation of competing cures to new diseases. Many elements from the French topical literature that accompanied Charles VIII's 1494–95 military campaign into Italy, for example, are eerily similar to what can be found in contemporary and subsequent narratives relating to the Americas: from the wonders of stunning riches and previously unknown fruits and vegetables, to enthusiastic accounts of the willingness of locals to be subdued and conquered, to the usual yet somewhat counterintuitive mixture of admiring otherness and wanting to annihilate it— *"toutefois on leur apprendra le train de france"* [however we will teach them the habits of France].[39] Reports from the Americas and the Franco-Italian Wars also share the obvious trait of referring to foreign locales; *foreignness*, as a discursive construct, helped shape topical literature formats, to the point where the seventeenth century's first newspapers "were almost entirely concerned with foreign news."[40]

As a matter of fact, directly relevant to the advent of periodicals, albeit usually not recognized as such, are mission newsletters. Mostly from the Americas, Asia, and the Middle East, they were heavily marketed, sometimes misleadingly, as

---

[35]   Michael A. Sherman, "Political Propaganda and Renaissance Culture: French Reactions to the League of Cambrai, 1509–1510," *Sixteenth Century Journal*, 8/2 (1977): pp. 97–128.

[36]   Augustin Redondo (ed.), *Les Discours sur le Sac de Rome de 1527* (Paris, 1999).

[37]   Sylvie Deswarte-Rosa, "L'expédition de Tunis (1535): images, interprétations, répercussions culturelles," in B. Bennassar and R. Sauzet (eds), *Chrétiens et Musulmans à la Renaissance* (Paris, 1998), pp. 75–132.

[38]   John Alden (ed.), *European Americana, vol. I: 1493–1600* (New York, 1980).

[39]   Cited in Jean-Pierre Séguin, "La découverte de l'Italie par les soldats de Charles VIII, 1494–1495," *Gazette des Beaux-Arts*, 6th ser., 58 (1961): pp. 131–2.

[40]   Brendan Dooley, "Postscript," in Dooley and Baron (eds), *Politics of Information*, p. 296.

providing both exotic and new locales, and the very latest news thereof.[41] By the late sixteenth century, Jesuit letters were massively distributed in accordance with their professed exoticism and incremental newness; witness the printer Michel Sonnius's title to his 1571 collection: *Recueil des plus fraisches lettres, escrittes des Indes Orientales, par ceux de la Compagnie du nom de Iesus, qui y font residence, & envoiées l'an 1568.69. & 70* [*Collection of the freshest letters written from the Oriental Indies by Jesuits living there, sent during the years 1568, 1569, and 1570*].[42] Exoticism, however, is only a superficial attribute of the *occasionnel* format. Séguin's recension of 517 *canards* from 1529 to 1631 lists many titles concerning the frightening Turks, yet the vast majority relates to crimes and criminals, natural calamities, dangerous beasts, miracles, monstrous births, and nearly 100 reports on "supernatural" atmospheric phenomena.[43] *Othering*, as a meta-category, rather more accurately accounts for the various descriptions and narratives on display. The news is staged as a spectacle alien to its public: calamities and crimes are set as extraordinarily catastrophic or evil, and criminals and beasts as heinous fiends. Local yet abnormal animals are described as chimeric amalgams, much like exotic species: Cuban fishes with the body of a shell-less turtle, the head of an ox, the agreeableness of dolphins, and the docility of an elephant,[44] rub shoulders with one outlandish calf sporting a lion's mane, the hindquarters of a horse, the belly of a stag, and the extra head of a dog, born on the 10th of May, 1569, in the village of Bellifontaine, two leagues away from Abbeville.[45]

Some of the material for the staging of "unprecedentedness" is demonstrably rehashed. Séguin lists a few cases of printers tampering with titles, dates, and names, in order to pass off old events as new.[46] Rehashing is, however, not always malicious: it can be an arbitrary, almost automatic function of print-culture processes. Pierre Desrey's updated *Chroniques* include, for the year 1509, a brief mention of seven Native North Americans brought back and exhibited in the city

---

[41]   See Vincent Masse, "Nouveautés et prophéties; les premières lettres missionnaires imprimés en langue française et le *Des Merveilles du Monde* de Guillaume Postel," in G. Poirier (ed.), *De l'Orient à la Huronie: écritures missionnaires et littérature d'édification aux XVIe et XVIIe siècles* (Laval, forthcoming).

[42]   Concerning the industry of the Jesuit Newsletters from Asia, see Donald F. Lach, *Asia in the Making of Europe*, vol.1 (Chicago, 1965), pp. 314–31 and 427–67.

[43]   Jean-Pierre Séguin, *L'Information en France avant le périodique: 517 canards imprimés entre 1529 et 1631* (Paris, 1964). On "prodigious" canards, see Jean Céard, *La nature et les prodiges: l'insolite au XVIe siècle* (Geneva, 1977), esp. ch. 19.

[44]   Pietro Martire d'Anghiera, *Extraict ou recueil des Isles nouvellement trouves en la grand mer Oceane* (Paris, 1532), p. 120v° (describing a manatee).

[45]   *Le vray pourtraict d'un Monstre nay d'une Vache le dixiesme iour de May, 1569, au village de Bellifontaine* (Paris, 1569).

[46]   Séguin, *L'Information*, pp. 18–20. A telling example: Marguerite de la Rivière, a young female criminal, was apparently executed first in 1596, then again in 1607, 1616, and 1623.

of Rouen.[47] Those "*sept hommes saulvages*" [seven wild men] are unexpectedly declared to be from an island alternately named "*Terre neufve*" [Newfoundland] and "*Orane en affrique*" [Oran in Africa], the latter referring to the Algerian city— which, of course, neither is an island, nor was "newly found" at the time. The confusion originates from a mistranslation of Johannes Multivallis's addition to Eusebius's *Chronicon*, which relates two *distinct* events of 1509: "*Oranum in Africa ab Hispanis capitur. Septem homines sylvestres ex ea insula (quae terra novi dicitur) Rothomagum adducti sunt cum cymba vestimentis & armis eorum*" [Oran in Africa was captured by the Spaniards. Seven Wild Men, of this island called Newfoundland, were brought to Rouen with their vessel clothes and weapons].[48] Yet Desrey's unfortunate translation and confusion of Multivallis is reiterated *verbatim* by the aforementioned *Quatriesme livre de la mer des hystoires* (1518).[49] In Jean Longis's *Mer des histoires* (1550), the conflation is complete; "Newfoundland," as a toponym, has vanished: "*Environ ce temps fut prins par les Portugalois en une terre nouvellement trouvee par eulx en lysle de Orane tirant vers affrique une maniere de gens sauvages*" [About the same time were captured by the Portuguese, in a newly found part of the island of Oran near Africa, a variety of savage people].[50] From one version to the next, yet apparently without additional sources being used, the anecdote is altered and expanded. Desrey ascribes the discovery of Newfoundland/Oran to Spaniards, and the capture of the Wild Men to Normans; but *La mer des histoires* (1550) attributes both the discovery and the capture to the Portuguese. Even more tellingly, Desrey's observation that "*Leurs viandes sont chairs rosties*" [They eat cooked meat][51] comes to be corrected according to contemporary stereotypes: "*Ilz mangeuent la chair crue*" [They eat uncooked meat].[52]

The setting of Oran as a newly discovered island is a proof *ab absurdo* that both "discovery" and "newness" are textual artifacts whose implementation is primarily governed by discursive constraints. New information obviously circulates, since mentions of newly found lands—even if falsely reported—move from one text to the next. Yet additions to chronicles primarily refer to alternate utterances. Pierre Bourdieu's term "circular circulation"—a vicious cycle of information built upon previously uttered statements, in which specific content is hard to put in, and harder yet to take out[53]—aptly, if anachronistically, describes the phenomenon of constant reiterations.

---

[47] Robert Gaguin, *Les grandes croniques* (Paris, 1514), p. 247v°.

[48] Eusebius, *Chronicon*, p. 172v°.

[49] *Le quatriesme livre de la mer des hystoires* (Paris, 1518), p. 27r°.

[50] *La mer des histoires* (Paris, 1550), p. 187.

[51] Gaguin, *Les grandes croniques*, p. 247v°. Compare "*Cibus eorum: carnes tostae*," in Eusebius, *Chronicon*, p. 172v°.

[52] *La mer des histoires* (Paris, 1550), p. 187.

[53] Pierre Bourdieu, *Sur la télévision* (Paris, 1996), p. 26.

On account of cognitive, doctrinal, and discursive constraints, the staging of "newness" and its "discovery" is highly stereotypical, that is to say *ritualized*. Eating uncooked meat—or human flesh—is such a highly-predictable characteristic in descriptions of cultural otherness, that its proclamation is called upon, as a component of a specific litany, even when it should have been irrelevant. Another term coined by Bourdieu is of value here: "*l'extraordinaire ordinaire*" [ordinary extraordinariness],[54] which designates discursive elements whose pertinence is based upon their professed unexpectedness, even while they are being constantly reported—such as monstrous births by *occasionnels*. Monstrous births are sometimes given a moral interpretation, or predictive properties, which are themselves highly predictable: they are signs of the decline of societal mores, or omens of great catastrophes to come. Critics of the *Amadis* phenomenon coined neologisms—"*amadiser*," "*amadigauliser*"—in order to mock the artificially bulked-up adventures of the eponymous hero, or even to scoff at stereotypical courtship procedures associated with the series.[55] By the same token, the discoveries of "New India" and "New Africa" share props, characters, and lines of dialogue, with that of the Americas.[56] It is much less demanding to deploy "discovery" and "newness" as a set of procedures—that is to say as discursive ceremonies—than as heuristic devices intent on shattering worldviews.

Conclusion

"Each literary text," according to Terence Cave, is a metaphorical "hapax": "even if inserted in a tradition or a set of rigorous conventions, it stresses its own difference,"[57] Regardless of Cave's attempt to connect singularity to the concept of literature, his observation also holds true in relation to all the texts mentioned so far. Ultimately, the staging of "discovery" and that of "newness" are textual and paratextual strategies toward claiming the *hapax*-status. Their display is, furthermore, concomitant with the advent of print culture and the invention of copyrights—the latter expressed, in early-modern France, by the granting of *privilèges*.[58] Evidently, early-modern Europe indeed featured many discoveries,

---

54   *Ibid.*, p. 19.

55   Simonin, "La disgrâce d'*Amadis*," pp. 224, 232.

56   See, e.g., Sebastian Münster's *Cosmographia*, first published in German in 1544.

57   From *hapax legomenon*—"once said"—which designates a word occurring only once in the written record of a language. See Terence Cave, *Pré-histoires: textes troublés au seuil de la modernité* (Genève, 1999), p. 12.

58   Not to be confused with an author's copyright or royalties, the *privilège* is the granting, by the royal chancery, sovereign courts, or various royal officials, of exclusive rights, to a publisher, in the printing of a new text or a new translation, within a particular locality and for a fixed term of years. Note that in the early decades of the sixteenth century, the seeking of *privilèges* was a voluntary act, which did not involve any vetting of the

inventions, and novelties, regardless of contentious specifics. Present early-modern studies—however less so than previously—are, furthermore, precisely insistent on texts deemed to be *hapaxes*: masterworks, prefigurations of modernity, nexuses of cultural intermingling and epistemic disturbances, and so on. Looking solely at "revolutionary" texts, however, easily gives rise to the impression that revolutionary contents are strongly correlated with claims of revolutionary content. I hope to have shown otherwise.

text for acceptability of its content. The process relied mainly on the criterion of newness: *privilèges* were normally restricted to texts which were being printed for the first time. See Febvre and Martin, *L'Apparition*, pp. 233–42 and 338–43; and Elizabeth Armstrong, *Before Copyright: The French Book-Privilege System, 1498–1526* (Cambridge, 1990), pp. 92–9.

Afterword
The Art of the Field

James Dougal Fleming

Among the many emblem-books of sixteenth-century Italy is the *Rime degli Academici occulti*.[1] Each of the "hidden academics" (members of a private learned society) is represented by a gnomic engraving, a Latin motto, a vernacular gloss, various verses, and an esoteric sobriquet: they are the Abstruso, Adombrato, Arcano, Chiuso, Desioso, Incognito, Notturno, Nubiloso, Offuscato, Oscuro, and Sepolto. Their interests range from rhetoric to metaphysics to agriculture, but in every case, their goal is to understand "all the worthiest matters that are allowed to the weak light of our Intellects," via communion with "superior and intellectual essences."[2] Ultimately—as the lover becomes like the beloved, and the moon like the sun[3]—the *academici occulti* hope to transform themselves into "the nature of God."[4]

Obviously, this is a highly Neoplatonic program. The Occulti are after esoteric knowledge, and knowledge as esoteric. For the essential forms of things "are hidden beneath matter and beneath accidents." This is true not only of elements and mixtures, but even of our own natures: "beneath obscure signs of movement, vegetable and sensible," one finds "our true excellence and most noble form."[5] In the celestial world, meanwhile, it is manifest (so to speak) that the truth is contained within "contrary motions and aspects, and obscure conjunctions."[6] As their over-arching emblem, the Occulti choose an image of a Silenus-figurine, more-or-less as described by Alcibiades in the *Symposium*: a satyr concealing "a most beautiful figure [*Idolo*] of a god or goddess," "in the hollow of its body [*nel*

---

[1]   *Rime de gli Academici occulti: con le loro imprese et discorsi* (Brescia, 1568). Translation with the assistance of Alessandra Capperdoni. See also Eveline Chayes, "Language of Words and Images in the *Rime degli academici occulti*, 1568: Reflections of the Pre-Conceptual?" in Lodi Nauta (ed.), *Language and Cultural Change: Aspects of the Study and Use of Language in the Later Middle Ages and the Renaissance* (Leuven, 2006), pp. 149–72.

[2]   *Rime*, sig. A1.

[3]   *Ibid.*, fols 119–124.

[4]   *Ibid.*, fol. 1.

[5]   *Ibid.*, pref., "*Discorso Interno al Sileno*," unsigned.

[6]   *Ibid.*

voto del corpo]." Protected from "air, dust, and struggle," the object of devotion—the truth—endures; precisely as a consequence of its "internal perfection."[7]

Now, the Silenus expresses the ancient view that special insight mandates reticence. Pythagoras, as the academic called the Arcano reminds us, wished that his disciples could even "unlearn to talk [*favellare*]."[8] "A secret is not a secret if it is not accompanied by SILENCE."[9] Accordingly, the Occulti must "hide the splendor of the high mysteries of truth from the eyes of plebeians," and from "the blind and distorted judgment of the crowd."[10] This is why the *Rime* itself takes emblematic form: the wise must never make their insights available to the unlearned, except under cover of "fables, parables, and love-stories."[11] Yet why make the insights available at all—and in a printed book, at that? Why publish the fact that there *is* a secret knowledge, possessed only by the esoteric authors themselves? To be sure, versions of this paradox are quite common at the early-modern juncture of print culture and *avant-garde* natural philosophy.[12] But that is what makes the *Rime* a publication of interest, for our purposes. How are we to understand its characteristic period tension between natural-philosophical esotericism, and technological manifestation?

The best answer seems to be that a publication like the *Rime* does not only posit secrecy as a communicative or social adjunct of advanced or powerful insights. Rather, it posits secrecy as an ontological and ineliminable *conjunct* of any such insight. For the highest truths, like God's appearing to Moses in a cloud, are always, and necessarily, "involved and hidden in obscurity."[13] Grasping the truth, therefore, *means* grasping that obscurity, in some kind of hieroglyphic *Gestalt*. "Men have always communicated among themselves what they conceive in their soul through letters," we are told; but "the Egyptian priests expressed many of their most hidden and profound thoughts through figures of natural bodies."[14] Emblematic rendering is as though *cognitively*, perhaps even epistemologically, requisite to the very nature of the "most hidden and profound" truths. This, too, is a fairly familiar attitude in the period.[15] It is as if these Renaissance Neoplatonists,

---

[7]   *Ibid.*

[8]   *Ibid.*, sig. E5ᵛ.

[9]   *Ibid.*, sig. E4ᵛ.

[10]   *Ibid.*, pref., "*Discorso.*"

[11]   *Ibid.*

[12]   See William Eamon, *Science and the Secrets of Nature: Books of Secrets in Medieval and Early Modern Culture* (Princeton, 1994); Pamela O. Long, *Openness, Secrecy, Authorship* (Baltimore, 2001); and Allison Kavey, *Books of Secrets: Natural Philosophy in England, 1550–1600* (Urbana, 2007).

[13]   *Rime*, pref., "*Discorso.*"

[14]   *Ibid.*

[15]   See Edgar Wind, *Pagan Mysteries in the Renaissance* (London, 1968); Don Cameron Allen, *Mysteriously Meant: The Rediscovery of Pagan Symbolism and Allegorical Interpretation in the Renaissance* (Baltimore, 1970); Peter M. Daly, *Literature in the Light*

freed from sober scholastic categories, and thrilled by the possibilities of their own esotericism, cannot quite think the latter through to its own termination. They cannot quite conceptualize, in other words, the transformation of natural secrets—those worthy of the name—into *mere* facts. To put it in the terms that began the current book: they do not yet have the hermeneutics of discovery.

But they are reaching toward it. The emblem devoted to the academician called the Incognito depicts a plough, which has run onto a hoard. The motto is "VETERES TELLURE RECLUDIT [it reveals treasures in the earth]." The commentary runs:

> There are many who distinguish (after Aristotle) two accidental causes, Fortune and Fate; the one occurs in those matters that are done by choice, the other in those that are done beyond it. Therefore, since Ploughing is an action proceeding from a man, who works by choice, they say that if one who Ploughs should discover treasure in the course of turning the earth, this discovery is Fortune, and the Ploughman is not the *per se* cause of finding the treasure.

This recalls Petrarch's traditional disdain for the fisherman who finds a gem inside of his catch.[16] It is the whole Aristotelian *de*-emphasis of discovery, as a function of ontological accidence, rather than substance. The commenting academician, however, has more to say:

> Through the effect of Fortune, or accidental cause, our academic INCOGNITO has wished to signify, instead, a profound mystery in the Art of the fields ... allegorically, he has the treasure, which was underground and hidden from other eyes, discovered by the Plough, as by the instrumental cause employed by the cultivator of the earth ... one sees how this praiseworthy man has created this Impresa in order to show other men that a diligent, laborious, and appropriate Cultivation, signified by the plough, is the reason why the earth produces most useful treasures for all human needs.[17]

The lucky ploughman, previously emblematic of the way things *aren't*, now becomes emblematic of the way things *are*. The plough, precisely as an unknowing and merely instrumental cause of "useful treasures," becomes the key to the whole wondrous productivity of the earth. And this because the category of fortune—the intervention into human choice by something that the human does not choose—locates substance *in the earth*, rather than in the human. *Precisely because*

---

of the Emblem (Toronto and London, 1998), pp. 42–58; and Daniel S. Russell, "Perceiving, Seeing and Meaning: Emblems and Approaches to Reading in Early Modern Culture," in Peter M. Daly and John Manning (eds), *Aspects of Renaissance and Baroque Symbol Theory, 1500–1700* (New York, 1999), pp. 77–92.

16   See Introduction, this volume, pp. 11–24.

17   *Rime*, fols 40–41.

he does not know what he is doing, the lucky ploughman indicates a decisive reconfiguration of the relationship between worldly forces and human purposes. This is to be a relationship, not of familiarity or belonging or fore-understanding; but of happening-upon, surprisingly encountering, being-interrupted, finding by chance. The Incognito and his glossator have just invented discovery.

Let us try to theorize this move, a little more fully than heretofore, by way of an overall conclusion to the current book. (Obviously, this is not to suggest any subsequent influence for the *Rime*.) Evidently, the Occulti do not *assume* the category or idea of discovery, in its hermeneutic function as the crucial gateway to knowledge, as part of their inheritance from classical or medieval intellectual traditions. For if they assumed it, they wouldn't have to invent it; and even if they did, the invention would (presumably) be much less tendentious and inchoate than the theorization they actually provide. A text like the *Rime* presents discovery, in other words, as an *emerging*, and as yet perplexing, function of the early-modern intellectual scene. We have seen considerable evidence for this basic point in the discussions above.

And what emerges? Grant an invention of discovery in the period; what is the actual *content* of the alleged invention? The Occulti, if we can continue to treat their ploughman as an interesting example, are locating the core productivity of human action in a *radical objectivity*: a worldly presence-at-hand (in Heidegger's phrase), over and against human purposes and expectations.[18] Indeed, they are *projecting* and *idealizing* that objectivity, precisely through their ontological and epistemological rehabilitation of the fortunate ploughman. The very non-intentionality of the ploughman's find—the way he comes upon, without anticipating, the treasure—makes his find exemplary for "the art of the fields." By extension and analogy, the unanticipated treasure becomes emblematic for the desideratum of natural-philosophical progress. A fortunate discovery, precisely as such, means that a fact of the world, as though of its own accord, has *announced itself*. This produces objective self-evidence, via subjective alienation.

We begin to see here some of the lineaments of modern natural science. The scientist, at least according to his popular profile, submits to subjective alienation for the sake of objective self-evidence. He seeks to create the conditions under which a fact of the world can announce itself; rather than the conditions under which the scientist's own intentionality can talk. To be sure, the scientist has to go out *looking for* the facts; and that means having some provisional idea of where (in what field of data) and how (by what instrumental tools) they may be found. Yet it would make no difference to the validity or interest of a given fact if it just dropped into the researcher's lap. Indeed, if and when a fact *is* found, within a gathered set of data, what makes it a matter of more or less significance is just whether or not it presents itself as some kind of *surprise*. True, under what

---

[18] See Martin Heidegger, *Being and Time*, trans. John Macquarie and Edward Robinson (New York, 1962), pp. 79–107, 121–36.

Kuhn calls normal science, big surprises may be rare.[19] But scientific revolutions occur when somebody finds something astonishing. Running onto treasure, for the modern natural scientist as for the Incognito, is the best conceivable reason for bringing the plough to a halt.

The invention of discovery, then, is highly germane to the early-modern emergence of modern natural science. Specifically, and as I have suggested already, it is *hermeneutically* germane. This is as distinct from, and running deeper than, any sociological or philological analysis. Discovery co-ordinates a proper scientific relationship—or what is popularly supposed to be a proper scientific relationship—between the interpretation of data, and the facts that emerge from that data. A fact is supposed to be something that is out there in the world; not in the investigator of the world. Discovery, with its correlate of surprise, secures precisely this worldly objectivity to the putative fact. The hermeneutics of discovery, in short, underwrites scientific objectivism.

Needless to say, the question of objectivism—"theory-free observation," presenting for analysis the world just as it is—has agitated philosophy of science from Hume onwards. Discovery, as linked to objectivism, is implicated and involved in that debate. Indeed, it is possible that a theoretical focus on the hermeneutics of discovery provides a useful clarification of the relevant cognitive and/or epistemological issues. For the trope of discovery is supposed to discipline data-interpretation, in service of the facts. The objectivist wants to interpret data in order to reach the facts just as they are. That means *discovering* facts, through a process of data-interpretation that terminates itself precisely when the facts are reached. Facts require discovery. But discovery is validated—one knows that it has occurred—if and only if it actually leads to some facts. Discovery requires facts. It is very remarkable to think that scientific objectivism, the very archetype and authority of knowledge-production in modernity, consists in the incessant repetition and re-imposition of this patent circularity.

For current purposes, however, it is not necessary (or desirable) to follow up that critical thought. It is necessary only to point out the burden that it implies for students of the early-modern period, insofar as this is organized around the emergence of modern natural science. For the trope of discovery, if indeed it was canonized by emergent science for its own purposes, cannot then be deployed neutrally in the attempt to understand emergent science. That would be to *presuppose* science, in its hermeneutic dimension; not to examine it. If we assume discovery as an interpretative trope for our analyses of the early-modern scene, we are in fact projecting backward a certain product of that scene. We are accepting, uncritically, the early-modern invention; and reifying, willy-nilly, the Enlightenment sense of discovery, *a fortiori* of scientific objectivism, as normative for understanding.

---

[19]   See Thomas S. Kuhn, *The Structure of Scientific Revolutions*, 3rd ed. (Chicago, 1996).

Let us consider how this sort of error might be committed in the case of the *Rime*. (The example is pleasant because I am *not* aware of any scholarship that actually does what it describes.) Almost reflexively, one might approach the text's emblems—these hypostases of the hidden and the open—under a regime of encoding and decoding, figurative and literal. This, clearly, would be a regime consistent with the hermeneutics of discovery. Interpreting the *Rime*, accordingly, might mean designating the referential intentions that the Occulti have merely obscured or deferred with emblematic clothes. Alternatively, it might mean proposing that the book encodes no such intentions—and that the *Rime*, as a result, is merely aesthetic. Either such approach, however, would presuppose as normative, for the Occulti, exactly the hermeneutics and implied epistemology that we have seen them confusedly and modishly constructing. We simply do not know—but have good reason to doubt—that the Occulti think in terms of secret as against open, literal as against figurative, discovered as against merely given. The danger, therefore, of approaching their text through these objectivistic oppositions would be that we might *entirely fail to notice* whatever this extremely strange and yet characteristic period text actually has to teach us.

Of course, this hypothetical critique implicates itself. To raise the specter of anachronism is to suggest a countervailing standard of historical correctness; which can never be defended fully from notions of objectivity, seeing the past as it was, discovering its truth. That is to say, however, that the hermeneutics of discovery reaches *all the way down* in the modern or postmodern enterprises of interpretation and understanding. So vast is the horizon of modern natural science, so decisive the trope of discovery for its conception of factuality, that seeing beyond or without that trope is at best difficult, at worst impossible. As discussed in the Introduction to this book, even the revisionist turn from discovery to invention tends to assume discovery as normative. We postmodern students of the early-modern have a natural and perhaps ineluctable tendency to bring discovery *with* us, even as we try to go back behind it.

How, then, are we to proceed? We need to recognize that we are dealing with a manifestly hermeneutic problem. That is to say, the issue before us is not just how to understand a given historical object (difficult though that be); but how to understand, *kurz*. For the object in question—the early-modern emergence of natural science—itself entails certain fundamental yet questionable claims about what counts as understanding. Therefore, studying the early-modern period, with the emergence of modern natural science at its core, means doing hermeneutics: trying to understand understanding. This is not just something that can follow or be added to an historical grasp of the field. It is the way in which the historical field in question demands to be grasped. Early-modern studies, with regard to the emergence of modern natural science, is hermeneutic *by definition*.

Where understanding is the object, the *way to* that object must itself be thrown in question. For knowing how to achieve it—which disciplines and techniques are in, which out—would presuppose it. Interdisciplinarity, in a word, must rule early-modern intellectual history. This is consistent with the highly eclectic period origins of modern natural science in alchemy, theology (including exegesis), natural philosophy, craft and trade traditions, etc. To be sure, interdisciplinarity is a very general custom of postmodern academia—but one very often breached in the very observance. It is easy to marshal large numbers of disparate scholars under a vast and flapping banner—"the Renaissance," for example. But unless they have some common intellectual object, they have nothing much to talk about. The large and collective meeting of many minds dissolves into a collection of many small meetings. Meaningful interdisciplinarity must bring its members *to bear* on something, if the disciplines are to interact at all. This sort of common purpose is exactly what is enabled, and demanded, by a version of early-modern studies that places emergent science at its core.

Hermeneutic interdisciplinarity, in conclusion, is the art of the early-modern field. To study the period, with emergent science as its treasure, is to try to understand understanding, with no holds barred. This does not involve deconstructing scientific factuality, but *avoiding the foreclosure* of its inductive ground. At play on this ground, I have been arguing, are our very expectations and assumptions about what it is to understand, and how one goes about it. Nothing could be more hermeneutic than the emergent notion of discovery in early-modern European culture. Nothing could be more germane to the study of early-modern European culture than the interdisciplinary attempt to understand this emergence. Nothing could be more productive for our own understanding of understanding than the unique and powerful challenges presented by this transformative period invention.

Bibliography

Primary Sources

Accademia degli Occulti, *Rime de gli Academici occulti: con le loro imprese et discorsi* (Brescia, 1568).

Agrippa, Cornelius, *Three Books of Occult Philosophy*, trans. J.F. (London, 1651).

Apollodorus, *The Library of Greek Mythology*, trans. Robin Hard (Oxford: Oxford University Press, 1997).

Aquinas, St. Thomas, "The Letter of Thomas Aquinas *De occultis operibus naturae ad quendam militem ultramontanum*," trans. J.B. McAllister, in *The Collected Works of St. Thomas Aquinas, Electronic edition* (Charlottesville: InteLex Corporation, 1993), pp. 21–30.

Aristotle, *The Basic Works*, ed. and intro. Richard McKeon (New York: Random House, 1941).

——, *Poetics*, trans. James Hutton (New York: Norton, 1982).

Bacon, Francis, *The Works of Francis Bacon*, eds James Spedding, Robert Leslie Ellis, and Douglas Dennon Heath (14 vols, London: Longmans, 1858–74).

——, *The Major Works*, ed. Brian Vickers (Oxford: Oxford University Press, 1996).

——, *The Advancement of Learning*, ed. Michael Kiernan (Oxford: Clarendon Press, 2000).

Belleforest, François de, *L'Histoire universelle du monde, contentant l'entiere description & situation des quatre parties de la terre* (Paris: Gervais Mallot, 1570).

Benedetti, Giovanni Battista, *Diversarum speculationum mathematicarum et physicarum liber* (Turin, 1585).

Bostocke, Richard, *The Difference between the Auncient Physicke ... and the Latter Physicke* (London, 1585).

Brahe, Tycho, *Epistolarum astronomicarum libri* [Uraniborg, 1596], in *Opera Omnia* ed. J.L.E. Dreyer (Amsterdam: Swets & Zeitlinger, 1972).

British Library MSS Sloane 1627, 1744, 1800, and 2218.

Brome, Richard, *The English Moore; or the Mock-Marriage*, ed. Sara Jayne Steen (Columbia: University of Missouri Press, 1983).

Bruno, Giordano, *Opera latine conscripta* (Stuttgart-Bad Cannstatt: Fromman Verlag, 1962).

——, *The Ash Wednesday Supper*, trans. S.L. Jaki (The Hague: Mouton, 1975).

——, *Dialoghi filosofici italiani*, 2nd ed. (Milano: A. Mondadori, 2000).

Campanella, Tommaso, *Poëtica*, in Luigi Firpo (ed.), *Tutte Le Opere di Tommaso Campanella*, vol. 1 (Verona: Mondadori, 1954).

——, *Del Senso delle Cose e della Magia*, ed. Germana Ernst (Bari: Laterza, 2007).

Canons and Decrees of the Council of Trent (London, 1687).

Case, John, *Lapis Philosophicus seu commentarius in 8 libris* (Oxford, 1599).

Cavendish, Margaret, *The Blazing World and Other Writings*, ed. Kate Lilley (New York: Penguin, 1992).

Copernicus, Nicholas, *On the Revolutions*, trans. Edward Rosen (Baltimore: Johns Hopkins University Press, 1992).

Cusa, Nicolaus de, *De docta ignorantia/Die belehrte Unwissenheit* (Hamburg, 1999).

Dekker, Thomas, *The Dramatic Works*, ed. Fredson Bowers, vol. 4 (Cambridge: Cambridge University Press, 1961).

Descartes, René, *The Philosophical Writings of Descartes*, eds John Cottingham, Robert Stoothoff, and Dugald Murdoch (3 vols, Cambridge: Cambridge University Press, 1991).

——, *The World and Other Writings*, ed. Stephen Gaukroger (Cambridge: Cambridge University Press, 1998).

Description des terres troves de nostre temps (Lyon, 1559).

Donne, John, *The Variorum Edition of the Poetry of John Donne*, vol. 6, ed. Gary Stringer (Bloomington: Indiana University Press, 1995).

Dorn, Gerhard, *Clavis Totius Philosophicae Chymisticae* (Lyon, 1566).

——, *Commentaria in Archidoxorum Libros X, D. Doctoris Theophrasti Paracelsi* (Frankfurt, 1584).

Eliot, John, *A Grammar of the Massachusetts Indian Language by John Eliot. A New Edition: With Notes and Observations by Peter S. Du Ponceau, L.L.D. and an Introduction and Supplementary Observations by John Pickering* (Boston: Phelps and Farnham, 1822).

Erastus, Thomas, *De Occultis Pharmacorum Potestatibus* (Basil, 1574).

Eusebius of Caesarea, *Eusebii Caesariesis Episcopi Chronicon ... Ad quem & Prosper & Matthaeus Palmerius / & Matthias Palmerius / demum & Ioannes Multivallis complura quae ad haec usque tempora subsecuta sunt adiecere* (Paris: H. Stephanum, 1512).

Fernel, Jean, *On the Hidden Causes of Things: Forms, Souls and Occult Diseases in Renaissance Medicine*, trans. John Forrester, eds Forrester and John Henry (Leiden: Brill, 2004).

Ficino, Marsilio, *Marsilio Ficino and the Phaedran Charioteer*, trans and ed. Michael B. Allen (Berkeley: University of California Press, 1981).

——, *Three Books on Life*, trans and eds Carol V. Kaske and John R. Clark (New York: Medieval and Renaissance Texts and Studies, 1989).

Fletcher, John, *Monsieur Thomas* (London, 1639).

——, and Francis Beaumont, *Comedies and Tragedies Written by Francis Beaumont and John Fletcher* (London, 1647).

Fontaine, Charles, *Les nouvelles, & Antiques merveilles* (Paris, 1554).

Foxe, John, *Eicasmi seu meditationes in sacram Apocalypsin* (London, 1587).

——, *Actes and Monuments ... The Variorum Edition* [online], vers. 1.1, http://www.hrionline.shef.ac.uk/foxe/ (Sheffield: hriOnline, 2006).

Fracastorius, Hieronymus, *In libros de Sympathia et Antipathia rerum*, in *Opera* (Lyon, 1591).

Fuller, Thomas, *The Holy State* (London, 1642).

——, *A Sermon of Reformation* (London, 1643).

——, *Truth Maintained* (London, 1643).

——, *A Pisgah-Sight of Palestine* (London, 1650).

Gaguin, Robert, *Les grandes croniques*, trans. Nicole de la Chesnaye (Paris, 1514).

Galilei, Galileo, *Sidereus Nuncius*, in *Le Opere, Edizione Nazionale* (20 vols, Florence: G. Barbera, 1929–39).

——, *Sidereus Nuncius or the Starry Messenger*, trans. and intro. Albert Van Helden (Chicago: University of Chicago Press, 1989).

——, *Galileo on the World Systems*, trans. and ed. Maurice A. Finocchiaro (Berkeley: University of California Press, 1997).

Gasser, Achille Pirminius, *Brief recueil de toutes Chroniques & Hystoires ... faictz & advenues depuis le commencement du monde isques au temps present, an mil cinq cens trente & quattre* (Anvers: Martin Lempereur, 1534).

Gregory of Nyssa, "Letter to Theodosius concerning the Belly-Myther," in Rowan A. Greer and Margaret M. Mitchell (trans.), *The "Belly-Myther" of Endor: Interpretations of 1 Kingdoms 28 in the Early Church* (Atlanta: SBL, 2007), pp. 166–78.

Hall, Joseph, *Contemplations on the Old Testament*, in *Bishop Hall's Works*, vol. 1 (Oxford: D.A. Talboys, 1837).

Harpsfield, Nicholas, *Dialogi sex contra summi pontificatus, monasticae vitae, sanctorum sacrarum imaginum oppugnatores, et pseudomartyres* (Antwerp, 1566).

Harriot, Thomas, *Artis analyticae praxis ad aequationes algebraicas nova, expedita, et generali methodo, resoluendas: tractatus* (London, 1631).

——, *A Briefe and True Report of the New Found Land of Virginia* (New York: Dover, 1972).

Herberay des Essarts, Nicolas d', *Le Cinqiesme livre d'Amadis de Gaule* (Paris, 1550).

Herbert, George, "Vanity (1)," in John Tobin (ed.), *George Herbert: The Complete English Poems* (London, 1991), pp. 78–9.

Homer, *The Iliad*, trans. Richard Lattimore (Chicago: University of Chicago Press, 1961).

Howell, James, *Instructions for forreine travel* (London, 1642).

Hutton, Henry, *Follies Anatomie, Or Satyres and Satyricall Epigrams with a Compendious History of Ixion's Wheele* (London, 1619).

James VI, *Daemonologie* (Edinburgh, 1597).

Kepler, Johannes, *Astronomia nova*, (Prague, 1609).

——, *Apologia Tychonis contra Ursum*, in *Opera omnia*, ed. C. Frisch (8 vols, Frankfurt am Main: Heyder und Zimmer, 1858–71).

——, *Gesammelte Werke*, eds Walther van Dyck and Max Caspar (20 vols, Munich: C.H. Beck, 1938–).

——, *Kepler's Conversation with the Starry Messenger*, trans. Edward Rosen (New York: Johnson Reprint, 1965).

——, *Somnium: The Dream, or Posthumous Work on Lunar Astronomy*, trans. Edward Rosen (Madison: University of Wisconsin Press, 1967).

——, *The Birth of History and Philosophy of Science: Kepler's "A Defence of Tycho against Ursus,"* trans. N. Jardine (Cambridge: Cambridge University Press, 1984).

——, *New Astronomy*, trans. William H. Donahue (Cambridge: Cambridge University Press, 1992).

——, *Optics: Paralipomena to Witelo and the Optical Part of Astronomy*, trans. William H. Donahue (Santa Fe: Green Lion Press, 2000).

Kircher, Athanasius, *Magneticum Naturae* (Rome, 1667).

Kyd, Thomas, *The First Part of Jeronimo* (London, 1605).

Le Gendre, Jean, *Tier livre de la fleur et mer des hystoires plus celebres & memorables advenues tant en l'Asie & Affricque qu'en l'Europpe, nouvellement recuillies & ordonnées selon la progression des temps & années, par Iehan le Gendre Aurelianoys, Mathematicien, commençant en l'an mil cinq cens trente & cinq, &continuant iusques en l'an mil cinq cens cinquante & ung* (Paris: Jean Longis, 1550).

Lederer, John, *The discoveries of John Lederer in three several marches from Virginia to the west of Carolina and other parts of the continent* (London, 1672).

Lemnius, Levinus, *The Secret Miracles of Nature in Four Books* (London, 1658).

Luther, Martin, *Lectures on Genesis. Chapters 1–5*, in *Luther's Works*, vol. 1, ed. Jaroslav Pelikan (Saint Louis: Concordia, 1958).

Magini, Giovanni Antonio, *Novae coelestium orbium theoricae* (Venice, 1589).

——, *Ephemerides Coelestium motuum ... secundum Copernici observationes ... supputatae* (Venice, 1609).

Marston, John, *The Selected Plays of John Marston*, eds Macdonald P. Jackson and Michael Neill (Cambridge: Cambridge University Press, 1986).

Martire d'Anghiera, Pietro, *Extraict ou recueil des Isles nouvellement trouves en la grand mer Oceane* (Paris: Simon de Colines, 1532).

Massinger, Philip, *The Parliament of Love*, in *The Dramatic Works of Massinger and Ford*, ed. Hartley Coleridge (London: E. Moxon, 1848).

Mästlin, Michael, *Ephemerides novae* (Tübingen, 1580).

——, *De astronomiae hypothesibus* (Heidelberg, 1582).

[*Mer des histoires*], *Le quatriesme livre de la mer des hystoires et cronicques de France* (Paris, 1518).

[*Mer des histoires*], *Le premier volume de la mer des histoires, auquel et le second ensuyvant est contenu tant du vieil testament que du nouveau toutes les hystoires, actes et faictz dignes de mémoire puis la création du monde jusques en l'an mil cinq cens. VL.* (Paris: Galliot du pré, 1536).

[*Mer des histoires*], *Le premier volume de la mer des histoires, auquel et le second ensuyvant est contenu tant du vieil testament que du nouveau toutes les hystoires, actes et faictz dignes de mémoire puis la création du monde jusques en l'an mil cinq cens XLIII selon la cotte et datte des ans* (Paris: C. Langelier, 1543).

[*Mer des histoires*], *Le premier volume de la mer des histoires. Auquel & le second ensuyvant Est contenu tant du vieil testament que du nouveau toutes les hystoires / Actes / & faictz dignes de memoire / puis la creation du monde iusques en l'an Mil cinq cens. l. selon la cotte & datte des ans. Ensemble les choses faictes & advenues en France de puis l'an mil.ccccc xliii. iusques en ceste presente annee* (Paris: Jean Longis, 1550).

de Metz, Gossuin, *Mirouer du monde nouvellement imprime* (Geneva, 1517).

Milton, John, *Of Reformation* (London, 1641).

——, *Eikonoklastes*, in *Complete Prose Works of John Milton*, ed. Merritt Y. Hughes, vol. 3 (New Haven: Yale University Press, 1962).

More, Thomas, *The Complete Works*, eds Louis L. Martz, Richard S. Sylvester, and Clarence H. Miller (15 vols, New Haven: Yale University Press, 1963–97).

Olmo, Giovanni, *De Occultis in Re Medica Proprietatibus* (Brescia, 1597).

Pappus of Alexandria, *Pappi Alexandrini Mathematicae collectiones*, trans. Federico Commandino (Pisa, 1588).

Penuen, Sands, *Ambitions Scourge* (London, 1611).

Persio, Antonio, *Trattato dell'Ingegno dell'Huomo*, ed. Luciano Artese (Pisa: Istituti editoriali poligrafici internazionali, 1999).

Peucer, Kaspar, *Hypotheses Astronomicae* (Wittenberg, 1571).

Polo, Marco, *La Description geographique des provinces & villes plus fameuses de l'Inde Orientale, meurs, loix & coustumes des habitans d'icelles, mesmement de ce qui est soubz la dominiation du grand Cham Empereur des Tartares* (Paris: E. Grouleau / J. Longis / V. Sertenas, 1556).

Porta, Giambattista della, *Natural magick by John Baptista Porta, a Neapolitane* (London, 1669).

Pseudo-Dionysius the Areopagite, *Dionysiaca; recueil donnant l'ensemble des traductions latines* (Bruges: Desclée, 1937).

——, *The Complete Works*, trans. Colm Luibheid (New York: Paulist Press, 1987).

Raleigh, Sir Walter, *The History of the World*, ed. C.A. Patrides (Philadelphia: Temple University Press, 1971).

Ross, Alexander, *The philosophicall touch-stone, or, Observations upon Sir Kenelm Digbie's Discourses of the nature of bodies and of the reasonable soul* (London, 1645).

——, *The new planet no planet, or, The earth no wandring star, except in the wandring heads of Galileans* (London, 1646).

——, *Arcana microcosmi, or, The hid secrets of man's body discovered* (London, 1652).

Saltmarsh, John, *Examinations, or, a Discovery of some Dangerous Positions* (London, 1643).

Schön, Caspar, *Disputatio Physica qua Qualitates Occultas* (Wittenberg, 1665).

Sennert, Daniel, *De occultis qualitatibus*, in *Hypomnemata Physica*, bk 2 (Frankfurt am Main, 1636).

Tyndale, William, "A Prologue into the Second Book of Moses Called Exodus," in Thomas Russell (ed.), *The Works of the English Reformers: William Tyndale and John Frith*, vol. 1 (London: Ebenezer Palmer, 1831).

——, *An Answere vnto Sir Thomas Mores Dialoge*, in Anne M. O'Donnell and Jared Wicks (eds), *The Independent Works of William Tyndale*, vol. 3 (Washington, DC: Catholic University of America Press, 2000).

——, *The Obedience of a Christian Man*, ed. David Daniell (London: Penguin, 2000).

Voltaire, *Letters on England* [*Lettres philosophiques*], trans. Leonard Tancock (New York, 1980), pp. 57–61.

Le vray pourtraict d'un Monstre nay d'une Vache le dixiesme iour de May, 1569, au village de Bellifontaine (Paris, 1569).

Webster, John, *Three Plays*, intro. David C. Gunby (London, 1992).

Williams, Roger, *A Key into the Language of America (1643)* (Detroit: Wayne State Press, 1973).

Wright, Thomas, *The Passions of the Minde in Generall* (London, 1604).

Secondary Sources

Alden, John (ed.), *European Americana, vol. I: 1493–1600* (New York: Readex, 1980).

Alexander, Amir, "The Imperialist Space of Elizabethan Mathematics," *Studies in History and Philosophy of Science*, 26/4 (1995): pp. 559–91.

Allen, Don Cameron, "John Donne's Knowledge of Renaissance Medicine," *The Journal of English and Germanic Philosophy*, 52/3 (1943): pp. 322–42.

——, *Mysteriously Meant: The Rediscovery of Pagan Symbolism and Allegorical Interpretation in the Renaissance* (Baltimore: Johns Hopkins University Press, 1970).

Appelbaum, Robert. "Anti-geography," *Early Modern Literary Studies*, 4/2 (1998), http://purl.oclc.org/emls/04-2/appeanti.htm.

Armstrong, Elizabeth, *Before Copyright: The French Book-Privilege System, 1498–1526* (Cambridge: Cambridge University Press, 1990).

Badaloni, Nicola, "Sulla Costruzione e la Conservazione della Vita in Bernardino Telesio," *Studi Storici*, 30 (1989): pp. 25–42.

Bagchi, David and David C. Steinmetz (eds), *The Cambridge Companion to Reformation Theology* (Cambridge and New York: Cambridge University Press, 2004).

Baker, David Weil, *Divulging Utopia: Radical Humanism in Sixteenth-Century England* (Amherst: University of Massachusetts Press, 1999).

Barkan, Leonard, *Unearthing the Past: Archaeology and Aesthetics in the Making of Renaissance Culture* (New Haven: Yale University Press, 1999).

Barker, P. and B.R. Goldstein, "Theological Foundations of Kepler's Astronomy," *Osiris*, 16 (2001): pp. 88–113.

Barnes, Jonathan, "Aristotle's Theory of Demonstration," in Jonathan Barnes, Malcolm Schofield, and Richard Sorabji (eds), *Articles on Aristotle, 1: Science* (London: Duckworth, 1975), pp. 65–87.

Barrera-Osorio, Antonio. *Experiencing Nature: The Spanish American Empire and the Early Scientific Revolution* (Austin: University of Texas Press, 2006).

Barthelemy, Anthony Gerard, *Black Face Maligned Race: The Representation of Blacks in English Drama from Shakespeare to Southerne* (Baton Rouge: Louisiana State University Press, 1987).

Batho, Gordon R., "Thomas Harriot's Manuscripts," in Fox (ed.), *Thomas Harriot: An Elizabethan Man of Science*, pp. 286–97.

Beer, A. and P. Beer (eds), *Kepler: Four Hundred Years. Proceedings of Conferences Held in Honour of Johannes Kepler* (Oxford: Pergamon, 1975).

Bellanger, C., J. Godechot, P. Guiral and F. Terrou (eds), *Histoire générale de la presse françaises* (5 vols, Paris: Presses Universitaires de France, 1969–76).

Bellette, Anthony, "Art and Imitation in Donne's *Anniversaries*," *SEL*, 15/1 (1975): pp. 83–96.

Benson, Pamela, *The Invention of the Renaissance Woman: The Challenge of Female Independence in the Literature and Thought of Italy and England* (University Park: Pennsylvania State University Press, 1992).

Betteridge, Thomas, "From Prophetic to Apocalyptic: John Foxe and the Writing of History," in Loades (ed.), *John Foxe and the English Reformation*, pp. 210–32.

Betti, G.L., "Il copernicanesimo nello Studio di Bologna," in M. Bucciantini and M. Torrini (eds), *La diffusione del copernicanesimo in Italia. 1543–1610* (Florence: L.S. Olschki, 1997), pp. 67–82.

Biagioli, Mario, *Galileo's Instruments of Credit: Telescopes, Images, Secrecy* (Chicago: University of Chicago Press, 2006).

Blair, Ann, "Humanist Methods in Natural Philosophy: The Commonplace Book," *Journal for the History of Ideas*, 53/4 (1992): pp. 541–51.

——, *Theater of Nature: Jean Bodin and Renaissance Science* (Princeton: Princeton University Press, 1997).

Bleichmar, Daniela, Paula de Vos, Kristin Huffine, and Kevin Sheehan (eds), *Science in the Spanish and Portugese Empires, 1500–1800* (Stanford: Stanford University Press, 2009).

Bloch, Howard, *Etymologies and Genealogies: A Literary Anthropology of the French Middle Ages* (Chicago: University of Chicago Press, 1986).

——, "Genealogy as a Medieval Mental Structure and Textual Form," in H.U. Gumbrecht et al. (eds), *Grundriss der Romanischen Literaturen des Mittelalters*, 11/1 (Heidelberg: Carl Winter, 1986), pp. 135–6.

Bloomfield, Leonard, *Language* (New York: Holt, 1958).

Bönker-Vallon, A., "Bruno e Proclo: connessioni e differenze tra la matematica neoplatonica e quella bruniana," in Eugenio Canone (ed.), *La filosofia di Giordano Bruno. Problemi ermeneutici e storiografici* (Florence: L.S. Olschki, 2003), pp. 129–44.

Bono, James J., *The Word of God and the Languages of Man: Interpreting Nature in Early Modern Science and Medicine, Volume 1: Ficino to Descartes* (Madison: University of Wisconsin Press, 1995).

Boorstin, Daniel, *The Discoverers* (New York: Random House, 1983).

Boose, Lynda E., "'The Getting of a Lawful Race': Racial Discourse in Early Modern England and the Unrepresentable Black Woman," in Margo Hendricks and Patricia Parker (eds), *Women, 'Race' and Writing in the Early Modern Period* (London: Routledge, 1994), pp. 35–54.

Booth, Michael, "Thomas Harriot's Translations," *Yale Journal of Criticism*, 16/2 (2003): pp. 345–61.

Bostock, D., *Aristotle: Metaphysics Books Z and H* (Oxford: Clarendon Press, 1994).

Bourdieu, Pierre, *Sur la télévision* (Paris: Liber, 1996).

Bowden, William R., "The Bed Trick, 1603–1642: Its Mechanics and Effects," *Shakespeare Studies*, 5 (1969): pp. 112–23.

Bragdon, Kathleen, "Native Languages as Spoken and Written: Views from Southern New England," in Edward G. Gray and Norman Fiering (eds) *The Language Encounter in the Americas, 1492–1800* (Oxford and New York: Berghahn Books, 2000), pp. 173–88.

Brammall, Kathryn M., "Monstrous Metamorphosis: Nature, Morality and the Rhetoric of Monstrosity in Tudor England," *The Sixteenth Century Journal*, 27/1 (1996): pp. 3–21.

Brann, Noel L., *The Debate Over the Origins of Genius During the Italian Renaissance* (Boston: Brill, 2002).

Breitenberg, Mark, "The Flesh Made Word: Foxe's *Acts and Monuments*," *Renaissance and Reformation*, 25 (1989): pp. 381–407.

Bucciantini, M., *Galileo e Keplero. Filosofia, cosmologia e teologia nell'Età della Controriforma* (Turin: G. Einaudi, 2003).

Buszard, Laura Ann, "Constructional Polysemy and Mental Spaces in Potawatomi Discourse" (dissertation, dept. of Linguistics, University of California, Berkeley, 2003).

Cajori, Florian, "A Revaluation of Harriot's *Artis Analyticae Praxis*," *Isis*, 11 (1928): pp. 316–24.

Calhoon, Robert M., "On Political Moderation," *Journal of the Historical Society*, 6/2 (2006): pp. 275–95.

CamFreddoso, Alfred, "Medieval Aristotelianism and the Case against Secondary Causation in Nature," in Thomas V. Morris (ed.), *Divine and Human Action: Essays in the Metaphysics of Theism* (Ithaca: Cornell University Press, 1988), pp. 74–118.

Campbell, Mary Baine, *Wonder and Science: Imagining Worlds in Early Modern Europe* (Ithaca: Cornell University Press, 1999).

———, "Alternative Planet: Kepler's *Somnium* (1634) and the New World," in Claire Jowitt and Diane Watt (eds), *The Arts of 17th-Century Science: Representations of the Natural World in European and North American Culture* (Aldershot: Ashgate, 2002), pp. 232–49.

Cañizares-Esguerra, Jorge, *Nature, Empire, and Nation: Explorations of the History of Science in the Iberian World* (Stanford: Stanford University Press, 2007).

Canny, Nicholas, *The Origins of Empire* (Oxford: Oxford University Press, 1998).

Carey, John, *John Donne: Life, Mind, and Art* (New York: Oxford University Press, 1981).

Cave, Terence, *Pré-histoires: textes troublés au seuil de la modernité* (Genève: Droz, 1999).

Céard, Jean, *La nature et les prodiges: l'insolite au XVIe siècle* (Geneva: Droz, 1977).

Chapman, Alison A., "Now and Then: Sequencing the Sacred in Two Protestant Calendars," *Journal of Medieval and Early Modern Studies*, 33 (Winter 2003): pp. 92–123.

Charles, David, *Aristotle on Meaning and Essence* (Oxford: Oxford University Press, 2002).

Chartier, Roger, "Le périodique: les antécédents," in H.-J. Martin and R. Chartier (eds), *Histoire de l'édition française*, vol. 1 (Paris: Promodis, 1982), pp. 411–12.

Chayes, Eveline, "Language of Words and Images in the *Rime degli academici occulti*, 1568: Reflections of the Pre-Conceptual?" in Lodi Nauta (ed.), *Language and Cultural Change: Aspects of the Study and Use of Language in the Later Middle Ages and the Renaissance* (Leuven, 2006), pp. 149–72.

Cheyfitz, Eric, *The Poetics of Imperialism: Translation and Colonization from "The Tempest" to "Tarzan"* (New York: Oxford University Press, 1991).

Christianson, J.R., *On Tycho's Island: Tycho Brahe and His Assistants, 1570–1601* (Cambridge: Cambridge University Press, 2000).

Ciliberto, M. *La ruota del tempo* (Rome: Editori riuniti, 1987).

Civil, Pierre and Danielle Boillet (eds), *L'Actualité et sa mise en écriture aux XVe–XVIe et XVIIe siècles: Espagne, Italie, France et Portugal* (Paris: Presses Sorbonne nouvelle, 2005).

Claretta, G., "Lettere tre di Francesco Patrici a Giambattista Benedetti matematico del Duca di Savoia," in *Miscellanea di Storia Italiana*, vol. 1 (Turin: Fratelli Bocca, 1862), pp. 380–83.

Cloud, Random [Randall McLeod], "Where Angels Fear to Read," in Joe Bray, Miriam Handley, and Anne C. Henry (eds), *Ma(r)king the Text* (Aldershot: Ashgate, 2000), pp. 144–92.

Clucas, Stephen, "Thomas Harriot and the Field of Knowledge in the English Renaissance," in Fox (ed.), *Thomas Harriot: An Elizabethan Man of Science*, pp. 93–136.

Cohen, I.B. (ed.), *Puritanism and the Rise of Modern Science: The Merton Thesis* (New Brunswick: Rutgers University Press, 1990).

Collard, Franck, "Histoire de France en latin et histoire de France en langue vulgaire: la traduction du *Compendium de origine et gestis francorum* de Robert Gaguin au début du xvie siècle," in Y.-M. Bercé and Ph. Contamine (eds), *Histoires de France, historiens de la France* (Paris: Librairie Honoré Champion, 1994), pp. 91–118.

Cook, Amy, "Staging Nothing: Hamlet and Cognitive Science," *SubStance*, 35/2 (2006): pp. 83–99.

Cook, Harold, *Matters of Exchange: Commerce, Medicine, and Science in the Dutch Golden Age* (New Haven and London: Yale University Press, 2007).

Costa, Dennis, "Poetry and Gnosticism: The *poetica* of Tommaso Campanella," *Viator: Medieval and Renaissance Studies*, 15 (1984): pp. 405–18.

Coulson, Seana, *Semantic Leaps: Frame-Shifting and Conceptual Blending in Meaning Construction* (Cambridge: Cambridge University Press, 2006).

Crowley, John E., *The Invention of Comfort: Sensibilities and Design in Early-Modern Britain and Early America* (Baltimore: Johns Hopkins University Press, 2001).

Crystal, David (ed.), *The Cambridge Encyclopedia of Language* (Cambridge: Cambridge University Press, 1987).

Cumming, William P., *The Southeast in Early Maps*, 3rd ed. (Chapel Hill: University of North Carolina Press, 1998).

Cunnar, Eugene R., "Donne's 'Valediction: Forbidding Mourning' and the Golden Compasses of Alchemical Creation," in L. Frank (ed.), *Literature and the Occult: Essays in Comparative Literature* (Arlington: University of Texas at Arlington, 1977).

Daileader, Celia, *Racism, Misogyny, and the Othello Myth: Inter-racial Couples from Shakespeare to Spike Lee* (Cambridge: Cambridge University Press, 2005).

Daly, Peter M., *Literature in the Light of the Emblem* (Toronto and London: University of Toronto Press, 1998).

Dancygier, Barbara (ed.), *Language and Literature*, 15/1 (2006).

Daston, Lorraine, "Baconian Facts, Academic Civility, and the Pre-History of Objectivity," *Annals of Scholarship*, 8 (1991): pp. 337–63.

——, and Katherine Park, *Wonders and the Order of Nature, 1150–1750* (New York: Zone, 1998).

De Pace, A., *Le matematiche e il mondo. Ricerche su un dibattito in Italia nella seconda metà del Cinquecento* (Milan: FrancoAngeli, 1993).

de Vivo, Filippo, *Information and Communication in Venice: Rethinking Early Modern Politics* (Oxford: Oxford University Press, 2007).

Dear, Peter, *Discipline and Experience: The Mathematical Way in the Scientific Revolution* (Chicago: University of Chicago Press, 1995).

——, *The Intelligibility of Nature: How Science Makes Sense of the World* (Chicago and London: University of Chicago Press, 2006).

Delbourgo, James, and Nicholas Dew (eds), *Science and Empire in the Atlantic World* (New York: Routledge, 2008).

Derbyshire, John, *Unknown Quantity: A Real and Imaginary History of Algebra* (Washington, DC: Joseph Henry Press, 2006).

Des Chesne, Dennis, *Physiologia: Natural Philosophy in Late Aristotelian and Cartesian Thought* (Ithaca: Cornell University Press, 1996).

Desens, Marliss C., *The Bed-Trick in English Renaissance Drama: Explorations in Gender, Sexuality, and Power* (London: Associated University Presses, 1994).

Deswarte-Rosa, Sylvie, "L'expédition de Tunis (1535): images, interprétations, répercussions culturelles," in B. Bennassar and R. Sauzet (eds), *Chrétiens et Musulmans à la Renaissance* (Paris: H. Champion, 1998), pp. 75–132.

Di Bono, M., "L'astronomia copernicana nell'opera di Giovan Battista Benedetti," in *Cultura, scienze e tecniche nella Venezia del Cinquecento* (Venice: Istituto veneto di scienze, lettere ed arti, 1987), pp. 283–300.

——, *Le sfere omocentriche di Giovan Battista Amico nell'astronomia del Cinquecento* (Genoa: Consiglio nazionale delle ricerche, 1990).

Diringer, David, *The Story of the Aleph Beth* (New York: Thomas Yoseloff/World Jewish Congress, 1960).

Dobrzycki, J., "The Role of Observations in the Work of Copernicus," in A. Beer and A. Strand (eds), *Copernicus "Yesterday and Today"* (Oxford: Pergamon, 1975), pp. 27–35.

Docherty, Thomas, *John Donne, Undone* (New York: Methuen, 1986).

Doležel, Lubomir, *Heterocosmica: Fiction and Possible Worlds* (Baltimore: Johns Hopkins University Press, 1998).

Doniger, Wendy, *The Bedtrick: Tales of Sex and Masquerade* (Chicago: The University of Chicago Press, 2000).

Dooley, Brendan and S.A. Baron (eds), *The Politics of Information in Early Modern Europe* (London: Routledge, 2001).

Ducros, Franc, *Tommaso Campanella poète* (Paris: Presses Universitaires de France, 1969).

Eamon, William, "From the Secrets of Nature to Public Knowledge," in D.C. Lindberg and R.S. Westman (eds), *Reappraisals of the Scientific Revolution* (Cambridge, Cambridge University Press, 1990), pp. 333–66.

——, *Science and the Secrets of Nature: Books of Secrets in Medieval and Early Modern Culture* (Princeton: Princeton University Press, 1994).

Eco, Umberto, "Small Worlds," *Versus: Quaderni di Studi Semiotici*, 52/3 (1989): pp. 53–70.

Eisenstein, Elizabeth, *The Printing Press as an Agent of Change* (Cambridge: Cambridge University Press, 1980).

Elsner, Jaś and Joan-Pau Rubiés (eds), *Voyages and Visions: Towards a Cultural History of Travel* (London: Reaktion, 1999).

Emison, Patricia, *Creating the "Divine" Artist from Dante to Michelangelo* (Boston: Brill, 2004).

Escobedo, Andrew, *Nationalism and Historical Loss in Renaissance England: Foxe, Dee, Spenser, Milton* (Ithaca: Cornell University Press, 2004).

Fallon, Stephen, *Milton among the Philosophers* (Ithaca: Cornell University Press, 1991).

Fattori, M. and Bianchi, M. (eds), *Spiritus. IV Colloquio Internazionale del Lessico Intellettuale Europeo* (Roma: Edizioni dell'Ateneo, 1984).

Fauconnier, Gilles, and Mark Turner, *The Way We Think: Conceptual Blending and the Mind's Hidden Complexities* (New York: Basic Books, 2002).

Febvre, Lucien and Henri-Jean Martin, *L'apparition du livre* (Paris: A. Michel, 1958).

Field, J.V., *Kepler's Geometrical Cosmology* (Chicago: University of Chicago Press, 1988).

——, *The Invention of Infinity: Mathematics and Art in the Renaissance* (Oxford and New York: Oxford University Press, 1997).

Findlen, Paula, *Possessing Nature: Museums, Collecting, and Scientific Culture in Early Modern Italy* (Berkeley: University of California Press, 1994).

—— (ed.), *Athanasius Kircher: The Last Man who Knew Everything* (New York. Routledge, 2004).

Finocchiaro, Maurice A. (trans. and ed.), *Galileo on the World Systems* (Berkeley: University of California Press, 1997).

Firth, Katharine R., *The Apocalyptic Tradition in Reformation Britain, 1530–1645* (Oxford: Oxford University Press, 1979).

Fisher, Philip, *Wonder, Rainbows, and the Aesthetics of Rare Experiences* (Cambridge: Harvard University Press, 2003).

Fleming, J.D., "Making Sense of Science and the Literal," in Killeen and Forshaw (eds), *The Word and the World*, pp. 45–60.

——, *Milton's Secrecy and Philosophical Hermeneutics* (Aldershot: Ashgate, 2008).

Fletcher, Angus, "Living Magnets, Paracelsian Corpses, and the Psychology of Grace in Donne's Religious Verse," *ELH*, 72/1 (2005): pp. 1–22.

Force, J.E., and R. Popkin (eds), *Newton and Religion: Context, Nature and Influence* (Dordrecht and Boston: Kluwer, 1999).

Fox, Robert (ed.), *Thomas Harriot: An Elizabethan Man of Science* (Aldershot: Ashgate, 2000).

Freccero, John, "Donne's 'Valediction: Forbidding Mourning'," *ELH*, 30 (1963): pp. 335–76.

Freedberg, David, *The Eye of the Lynx: Galileo, His Friends, and the Beginnings of Modern Natural History* (Chicago: University of Chicago Press, 2002).

Freeman, Thomas S., "Introduction: Over Their Dead Bodies: Concepts of Martyrdom in Late Medieval and Early Modern England," in Thomas S. Freeman and Thomas F. Mayer (eds), *Martyrs and Martyrdom in England c. 1400–1700* (Woodbridge: Boydell Press, 2007), pp. 1–34.

Friedman, John Block, "*Secretz de la Nature [Merveilles du Monde]*, Les," in J.B. Friedman and K. Mossler Figg (eds), *Trade, Travel, and Exploration in the Middle Ages: An Encyclopedia* (New York: Garland Pub, 2000), pp. 545–6.

Frisch, Andrea, *The Invention of the Eyewitness: Witnessing and Testimony in Early-Modern France* (Chapel Hill: North Carolina Studies in the Romance Languages and Literatures, 2004).

Froehlich, Karlfried, "Pseudo Dionysius and the Reformation of the Sixteenth Century," in Pseudo-Dionysius, *The Complete Works*, pp. 33–46.

Gadamer, Hans-Georg, *The Enigma of Health*, trans. Jason Gaiger and Nicholas Walker (Stanford: Stanford University Press, 1996).

——, *Truth and Method*, 2nd rev. ed., trans. rev. Joel Weinsheimer and Donald G. Marshall (New York: Continuum, 2004).

——, "The Universality of the Hermeneutic Problem," in Richard E. Palmer (ed.), *The Gadamer Reader* (Evanston: Northwestern University Press, 2007), pp. 72–88.

Gaukroger, Stephen, *Francis Bacon and the Transformation of Early-Modern Philosophy* (Cambridge: Cambridge University Press, 2001).

——, *The Emergence of a Scientific Culture: Science and the Shaping of Modernity, 1210–1685* (Oxford and New York: Oxford University Press, 2006).

Gavins, Joanna and Gerard Steen (eds), *Cognitive Poetics in Practice* (London: Routledge, 2003).

Genette, Gérard, *Seuils* (Paris: Seuil, 1987).

Gingerich, Owen, "Kepler's Place in Astronomy", in Beer and Beer (eds), *Kepler: Four Hundred Years*, pp. 261–8.

——, *The Eye of Heaven: Ptolemy, Copernicus, Kepler* (New York: American Physics Institute, 1993).

——, and R.S. Westman, "The Wittich Connection: Conflict and Priority in Late Sixteenth-Century Cosmology," special issue of the *Transactions of the American Philosophical Society*, 78/7 (1988).

——, and James Voelkel, "Tycho Brahe's Astronomical Campaign," *Journal for the History of Astronomy*, 29 (1998): pp. 1–34.

Goddard, Ives, "Comparative Algonquian," in Lyle Campbell and Marianne Mithun (eds), *The Languages of Native America: Historical and Comparative Assessment* (Austin: University of Texas Press, 1979), pp. 70–132.

Godwin, David, *Godwin's Cabalistic Encyclopedia: A Complete Guide to Cabalistic Magick*, 3rd ed. (St. Paul: Llewellyn Publications, 2003).

Goodblatt, Chanita, "From 'Tav' to the Cross: John Donne's Protestant Exegesis and Polemics," in Mary Arshagouni Papazian (ed.), *John Donne and the Protestant Reformation: New Perspectives* (Detroit: Wayne State University Press, 2003), pp. 221–46.

Gordon, Stuart, *The Encyclopaedia of Myths and Legends* (London: Headline Publishing, 1993).

Grafton, Anthony, "Humanism and Science in Rudolphine Prague: Kepler in Context," in *Defenders of the Text: The Traditions of Scholarship in an Age of Science, 1450–1800* (Cambridge: Harvard University Press, 1991), pp. 178–203.

——, with April Shelford and Nancy Siraisi, *New Worlds, Ancient Texts: The Power of Tradition and the Shock of Discovery* (Cambridge, MA and London: Belknap Press, 1992).

——, *What Was History? The Art of History in Early Modern Europe* (Cambridge: Cambridge University Press, 2007).

Granada, M.A., "Il rifiuto della distinzione tra potentia absoluta e potentia ordinata di Dio e l'affermazione dell'universo infinito in Giordano Bruno," *Rivista di storia della filosofia*, 49/3 (1994): pp. 495–532.

——, *El debate cosmológico en 1588. Bruno, Brahe, Rothmann, Ursus, Röslin* (Naples: Bibliopolis, 1996).

——, and D. Tessicini, "Copernicus and Fracastoro: The Dedicatory Letters to Pope Paul III, the History of Astronomy, and the Quest for Patronage," *Studies in History and Philosophy of Science*, 36 (2005): pp. 431–76.

——, "Synodis ex mundis," *Bruniana & Campanelliana*, 13/1 (2007): pp. 149–56.

——, "Kepler and Bruno on the Infinity of the Universe and of Solar Systems," *Journal for the History of Astronomy*, 39 (2008): pp. 469–95.

Grant, Edward, "In Defense of the Earth's Centrality and Immobility: Scholastic Reaction to Copernicanism in the Seventeenth Century," *Transactions of the American Philosophical Society*, n.s., 74/4 (1984): pp. 1–69.

Greenblatt, Stephen, *Renaissance Self-Fashioning: From More to Shakespeare* (Chicago: University of Chicago Press, 1980).

——, *Shakespearean Negotiations* (Berkeley: University of California Press, 1988).

Guglielmetti, Marziano, "Magia e Tecnica nella Poetica di Tommaso Campanella," *Rivista di Estetica*, 9 (1964): pp. 361–93.

Guibbory, Achsah, *Ceremony and Community from Herbert to Milton: Literature, Religion and Cultural Conflict in Seventeenth-Century England* (Cambridge: Cambridge University Press, 1998).

Hadot, Pierre, *The Veil of Isis: An Essay on the History of the Idea of Nature*, trans. Michael Chase (Cambridge: Belknap Press, 2006).

Hahn, Lewis Edwin (ed.), *The Philosophy of Hans-Georg Gadamer* (Chicago: Open Court, 1997).

Hall, Kim F., *Things of Darkness: Economies of Race and Gender in Early Modern England* (Ithaca: Cornell University Press, 1995).

Halliwell, Richard, *The Aesthetics of Mimesis* (Princeton: Princeton University Press, 2002).

Hallyn, Fernand, *The Poetic Structure of the World: Copernicus and Kepler*, trans. Donald M. Leslie (New York: Zone Books, 1990).

—— (ed.), *Metaphor and Analogy in the Sciences* (Dordrecht: Kluwer, 2000).

Hannaway, Owen, *The Chemists and the Word* (Baltimore: Johns Hopkins University Press, 1975).

Harbage, Alfred, *Annals of English Drama 975–1700* (London: Methuen, 1964).

Harris, Stephen J., "Mapping Jesuit Science: The Role of Travel in the Geography of Knowledge," in John W. O'Malley, Gauvin Alexander Bailey, Steven J. Harris, and T. Frank Kennedy (eds), *The Jesuits: Cultures, Science, and the Arts, 1540–1773* (Toronto: University of Toronto Press, 1999), pp. 212–40.

Harrison, Peter, *The Bible, Protestantism and the Rise of Natural Science* (Cambridge: Cambridge University Press, 1998).

——, "Voluntarism and Early Modern Science," *History of Science*, 40 (2002): pp. 63–89.

——, *The Fall of Man and the Foundations of Science* (Cambridge: Cambridge University Press, 2007).

——, "Reinterpreting Nature in Early Modern Europe: Natural Philosophy, Biblical Exegesis and the Contemplative Life," in Killeen and Forshaw (eds), *The Word and the World*, pp. 25–44.

Hart, F. Elizabeth, "Matter, System and Early Modern Studies: Outlines for a Materialist Linguistics," *Configurations*, 6 (1998): pp. 311–43.

——, "The Epistemology of Cognitive Literary Study," *Philosophy and Literature*, 25 (2001): pp. 314–34.

——, "Embodied Literature: A Cognitive-Poststructuralist Approach to Genre," in Alan Richardson and Ellen Spolsky (eds), *The Work of Fiction* (Aldershot: Ashgate, 2004), pp. 85–106.

Hawkes, David, *Idols of the Marketplace: Idolatry and Commodity Fetishism in English Literature, 1580–1680* (London: Palgrave, 2001).

Heidegger, Martin, *Being and Time*, trans. John Macquarie and Edward Robinson (New York: Harper, 1962).

Hellman, D.C., *The Comet of 1577* (New York: Columbia University Press, 1944).

Henry, John, "Occult Qualities and the Experimental Philosophy: Active Principles in Pre- Newtonian Matter Theory," *History of Science*, 24 (1986): pp. 335–81.

——, *The Scientific Revolution and the Origins of Modern Science*, 3rd ed. (New York: Palgrave, 2008).

Hermann, Paul, *Great Age of Discovery* (New York: Harper, 1958).

Hessayon, Ariel, and Nicholas Keene (eds), *Scripture and Scholarship in Early-Modern England* (Aldershot: Ashgate, 2006).

Hirsch, E.D., *Validity in Interpretation* (New Haven: Yale University of Press, 1967).

Hockett, Charles F., "What Algonquian is Really Like," *International Journal of American Linguistics*, 32/1 (Jan. 1966): pp. 59–73.

hooks, bell, *Black Looks: Race and Representation* (Boston: South End Press, 1992).

Howell, Kenneth J., *God's Two Books: Copernican Cosmology and Biblical Interpretation in Early Modern Science* (Notre Dame: University of Notre Dame Press, 2002).

Hunt, Lynn (ed), *The Invention of Pornography: Obscenity and the Origins of Modernity, 1500–1800* (New York: Zone, 1996).

Hutchison, Keith, "Dormitive Virtues, Scholastic Qualities, and the New Philosophies," *History of Science*, 29 (1991): pp. 245–78.

——, "What Happened to Occult Qualities in the Scientific Revolution?" in Peter Dear (ed.), *The Scientific Enterprise in Early Modern Europe: Readings from Isis* (Chicago and London: University of Chicago Press, 1997), pp. 86–106.

Hutson, Lorna, *The Invention of Suspicion: Law and Mimesis in Shakespeare and Renaissance Drama* (Oxford and New York: Oxford University Press, 2007).

Hutton, Sarah, "Iconisms, Enthusiasms and Origen: Henry More Reads the Bible," in Hessayon and Keene (eds), *Scripture and Scholarship in Early-Modern England*, pp. 192–207.

Iliffe, Rob, "Friendly Criticism: Richard Simon, John Locke, Isaac Newton and the Johannine Comma," in Hessayon and Keene (eds), *Scripture and Scholarship in Early-Modern England*, pp. 137–57.

Jacobs, Deborah, "Critical Imperialism and Renaissance Drama: The Case of 'The Roaring Girl'," in Dale M. Bauer and Susan Jaret McKinstry (eds), *Feminism, Bakhtin, and the Dialogic* (Albany: State University of New York Press, 1991), pp. 73–84.

Jardine, Lisa, *Francis Bacon: Discovery and the Art of Discourse* (Cambridge: Cambridge University Press, 1974).

Johnson, Mark, *The Body in the Mind: The Bodily Basis of Meaning, Imagination and Reason* (Chicago: University of Chicago Press, 1987).

Jones, Eldred, *Othello's Countrymen: The African in English Renaissance Drama, 1550–1688* (Oxford: Oxford University Press, 1965).

Jones, Matthew, *The Good Life in the Scientific Revolution* (Chicago: University of Chicago Press, 2006).

Jones, William, "Algonquian (Fox)," in Franz Boas (ed.), *Handbook of American Indian Languages, Part 1, Volume 2* (Bristol: Thoemmes Press, 2002), pp. 735–874.

Jonson, Ben, "Ben Jonson's Conversations with William Drummond of Hawthornden," in C.H. Hereford and P. Simpson (eds), *Ben Jonson* (Oxford: Clarendon Press, 1925).

Kavey, Allison, *Books of Secrets: Natural Philosophy in England, 1550–1600* (Urbana: University of Illinois Press, 2007).

Keene, Nicholas, "A Two-Edged Sword: Biblical Criticism and the New Testament Canon in Early Modern England," in Hessayon and Keene (eds), *Scripture and Scholarship in Early-Modern England*, pp. 94–115.

Keilen, Sean, *Vulgar Eloquence: On the Renaissance Invention of English Literature* (New Haven: Yale University Press, 2006).

Kelly, Donald R., "Between History and System," in Gianna Pomata and Nancy G. Siraisi (eds), *Historia: Empiricism and Erudition in Early Modern Europe* (Cambridge, MA and London: MIT Press, 2005), pp. 211–37.

Kennedy, William J., *Authorizing Petrarch* (Ithaca: Cornell University Press, 1994).

Killeen, Kevin, and Peter Forshaw (eds), *The Word and the World: Biblical Exegesis and Early-Modern Science* (New York: Palgrave, 2007).

King, John N., *Foxe's Book of Martyrs and Early Modern Print Culture* (Cambridge: Cambridge University Press, 2006).

Kollermann, Judith J., "The Centaur," in Malcolm South (ed.), *Mythical and Fabulous Creatures: A Source Book and Research Guide* (London: Greenwood Press, 1987), pp. 225–39.

Koyré, A., *From the Closed World to the Infinite Universe* (Baltimore: Johns Hopkins Press, 1957).

———, *The Astronomical Revolution: Copernicus-Kepler-Borelli*, trans. R.E.W. Maddison (Paris and London: Hermann / Methuen, 1973).

Kozhamthadam, Job, *The Discovery of Kepler's Laws: The Interaction of Science, Philosophy and Religion* (Notre Dame: University of Notre Dame Press, 1997).

Krause, Virginia, "Serializing the French *Amadis* in the 1540s," in M. Rothstein (ed.), *Charting Change in France around 1540* (Selinsgrove: Susquehanna University Press, 2006), pp. 40–62.

Kuhn, Thomas, *The Copernican Revolution* (Cambridge: Harvard University Press, 1957).

———, *The Structure of Scientific Revolutions*, 3rd ed. (Chicago: Chicago University Press, 1996).

Kuriyama, Constance Brown, *Christopher Marlowe: A Renaissance Life* (Ithaca: Cornell University Press, 2002).

Lach, Donald F., *Asia in the Making of Europe*, vol. 1 (Chicago: University of Chicago Press, 1965).

Lacouture, Jean, "L'histoire immédiate," in J. Le Goff and R. Chartier (eds), *La nouvelle histoire* (Paris: Retz-C.E.P.L., 1978), pp. 270–93.

Lakoff, George, and Rafael Nuñez, *Where Mathematics Comes From: How the Embodied Mind Brings Mathematics into Being* (New York: Basic Books, 2000).

Laqueur, Thomas, *Making Sex: Body and Gender from the Greeks to Freud* (Cambridge: Harvard University Press, 1990).

Lardellier, Pascal, *Les miroirs du paon: rites et rhétoriques politiques dans la France de l'Ancien Régime* (Paris: Champion, 2003).

Lattis, James M., *Between Copernicus and Galileo: Christoph Clavius and the Collapse of Ptolemaic Cosmology* (Chicago: University of Chicago Press, 1994), pp. 120–26.

Lincoln, Evelyn, *The Invention of the Italian Renaissance Printmaker* (New Haven: Yale University Press, 2000).

Lindberg, David C., and Ronald Numbers (eds), *God and Nature: Historical Essays on the Encounter Between Christianity and Science* (Berkeley: University of California Press, 1986).

Loades, David (ed.), *John Foxe and the English Reformation* (Aldershot: Scolar Press, 1997).

Lochman, Daniel T., "*Divus Dionysius*: Authority, Self, and Society in John Colet's Reading of the *Ecclesiastical Hierarchy*," *Journal of the History of Ideas*, 68/1 (2007).

Long, Pamela O., *Openness, Secrecy, Authorship* (Baltimore and London: Johns Hopkins University Press, 2001).

Lorente, Joaquin Martinez, "Possible World Theories and the Two Fictional Worlds of More's *Utopia*: How Much (and How) Can We Apply," *Sederi*, 6 (1996): pp. 117–23. Available at: sederi.org/docs/yearbooks/06/6_13_martinez.pdf.

Losse, Deborah N., *Sampling the Book: Renaissance Prologues and the French Conteurs* (Lewisburg: Bucknell University Press, 1994).

Lossky, Vladimir, *The Mystical Theology of the Eastern Church* (Crestwood: St. Vladimir's Seminary Press, 1976).

Lovejoy, A.O., *The Great Chain of Being* (Cambridge: Harvard University Press, 1936).

MacKenthun, Gesa, *Metaphors of Dispossession: American Beginnings and the Translation of Empire, 1492–1637* (Norman: University of Oklahoma Press, 1997).

Malieckal, Bindu, "'Hell's Perfect Character': The Black Woman as the Islamic Other in Fletcher's *The Knight of Malta*," *Essays in Arts and Sciences*, 28 (1999): pp. 53–66.

Malpas, Jeff, Ulrich Arnswald and Jens Kertscher (eds), *Gadamer's Century: Essays in Honor of Hans-Georg Gadamer* (Cambridge: MIT Press, 2002).

Mamiani, Maurizio, "To Twist the Meaning: Newton's *Regulae Philosophandi* Revisited," in Jed Z. Buchwald and I.B. Cohen (eds), *Isaac Newton's Natural Philosophy* (Cambridge: MIT Press, 2001), pp. 3–14.

Mandelbrote, Scot, "English Scholarship and the Greek Text of the Old Testament, 1620–1720: The Impact of Codex Alexandrinus," in Hessayon and Keene (eds), *Scripture and Scholarship in Early-Modern England*, pp. 74–93.

Markley, Robert, *Fallen Languages: Crises of Representation in Newtonian England, 1660–1740* (Ithaca: Cornell University Press, 1993).

Marmor, Andrei, *Interpretation and Legal Theory* (Oxford: Oxford University Press, 1992).

Marmura, Michael E., "Al-Ghazali," in Peter Adamson and R.C. Taylor (eds), *The Cambridge Companion to Arabic Philosophy* (Cambridge: Cambridge University Press, 2005), pp. 137–54.

Martens, Rhonda, *Kepler's Philosophy and the New Astronomy* (Princeton and Oxford: Princeton University Press, 2000).

Martin, Thomas L., *Poeisis and Possible Worlds: A Study in Modality and Literary Theory* (Toronto: University of Toronto Press, 2004).

Masse, Vincent, "Nouveautés et prophéties; les premières lettres missionnaires imprimés en langue française et le *Des Merveilles du Monde* de Guillaume Postel," in G. Poirier (ed.), *De l'Orient à la Huronie: écritures missionnaires et littérature d'édification aux XVIe et XVIIe siècles* (Laval, forthcoming).

Matthews, Steven, *Theology and Science in the Thought of Francis Bacon* (Aldershot: Ashgate, 2008).

McCanles, Michael, "The New Science and the *Via Negativa*: A Mystical Source for Baconian Empiricism," in Julie Robin Solomon and Catherine Gimelli Martin (eds), *Francis Bacon and the Refiguring of Early-Modern Thought* (Aldershot: Ashgate, 2005), pp. 45–68.

McKim, Donald (ed.), *The Cambridge Companion to Martin Luther* (Cambridge: University Press, 2003).

—— (ed.), *The Cambridge Companion to John Calvin* (Cambridge: Cambridge University Press, 2004).

McKnight, Stephen, *The Religious Foundations of Francis Bacon's Thought* (Columbia: University of Missouri Press, 2005).

McLuhan, Marshall, *The Gutenberg Galaxy: The Making of Typographic Man* (Toronto: University of Toronto Press, 1962).

McMullin, Ernan, "Galileo on Science and Scripture," in Peter Machamer (ed.), *The Cambridge Companion to Galileo* (Cambridge: Cambridge University Press, 1998), pp. 271–347.

Menn, Stephen P., "Metaphysics: God and Being," in A.S. McGrade (ed.), *The Cambridge Companion to Medieval Philosophy* (Cambridge: Cambridge University Press, 2003), pp. 147–70.

Menninger, Karl, *Number Words and Number Symbols: A Cultural History of Numbers*, trans. Paul Broneer (New York: Dover, 1969).

Meserve, Margaret, "News from Negroponte: Politics, Popular Opinion, and Information Exchange in the First Decade of the Italian Press," *Renaissance Quarterly*, 59/2 (2006): pp. 440–80.

Metzger, Bruce M., "Forgeries and Canonical Pseudepigrapha," *Journal of Biblical Literature*, 91/1 (1972): pp. 3–24.

Millen, Ron, "The Manifestation of Occult Qualities in the Scientific Revolution," in Margaret J. Osler and Paul Lawrence Farber (eds), *Religion, Science and Worldview: Essays in Honor of Richard S. Westfall* (Cambridge: Cambridge University Press, 1985), pp. 185–216.

Miller, Christopher, and George R. Hamell, "A New Perspective on Indian-White Contact: Cultural Symbols and Colonial Trade," *Journal of American History*, 73 (1986): pp. 311–28.

Mohamed, Feisal, *In the Anteroom of Divinity: The Reformation of the Angels from Colet to Milton* (Toronto: University of Toronto Press, 2008).

Monta, Susannah Brietz, *Martyrdom and Literature in Early Modern England* (Cambridge: Cambridge University Press, 2005).

Morgan, Jennifer L., "Some Could Suckle Over Their Shoulder: Male Travelers, Female Bodies, and the Gendering of Racial Ideology, 1500–1770," *The William and Mary Quarterly*, 3rd ser., 54/1 (1997): pp. 167–92 .

Moschovakis, Nicholas R., "Topicality and Conceptual Blending: *Titus Andronicus* and the Case of William Hacket," *College Literature*, 33/1 (2006): pp. 127–50.

Mosley, Adam, *Bearing the Heavens: Tycho Brahe and the Astronomical Community of the Late Sixteenth Century* (Cambridge: Cambridge University Press, 2006).

Moss, Jean Deitz, *Novelties in the Heavens: Rhetoric and Science in the Copernican Controversy* (Chicago: University of Chicago Press, 1993).

Muller, Richard A., and John L. Thompson (eds), *Biblical Interpretation in the Era of the Reformation* (Grand Rapids: W.B. Eerdmans, 1996).

Mullin, Harryette, "Optic White: Blackness and the Production of Whiteness," *Diacritics*, 24/2–3 (1994): pp. 71–89.

Mushin, Ilana, "Viewpoint Shifts in Narrative," in Jean-Pierre Koenig (ed.), *Discourse and Cognition: Bridging the Gap* (Stanford: CSLI, 1998), pp. 323–36.

Newman, William, *Atoms and Alchemy: Chymistry and the Experimental Origins of the Experimental Revolution* (Chicago: University of Chicago Press, 2006).

Newton, Arthur Percival, *The Great Age of Discovery* (London: University of London Press, 1932).

O'Gorman, Edmundo, *The Invention of America: An Inquiry into the Historical Nature of the New World and the Meaning of Its History* (Bloomington: Indiana University Press, 1961).

Oliphant, E.H.C., *The Plays of Beaumont and Fletcher: An Attempt to Determine Their Respective Shares and the Shares of Others* (New Haven: Yale University Press, 1927).

Olsen, Palle J., "Was John Foxe a Millenarian?" *Journal of Ecclesiastical History*, 45 (October 1994): pp. 600–624.

Omodeo, Pietro D., "La Stravagantographia di un 'filosofo stravagante'," *Bruniana & Campanelliana*, 14/1 (2008): pp. 11–23.

——, "La cosmologia infinitistica di Giovanni Battista Benedetti," *Bruniana & Campanelliana*, 16/1 (2009): pp. 149–58.

Parry, Graham, *The Trophies of Time: English Antiquarians of the Seventeenth Century* (Oxford: Oxford University Press, 1995).

Patrides, C.A., *Premises and Motifs in Renaissance Thought and Literature* (Princeton: Princeton University Press, 1982).

Pelikan, Jaroslav, *The Christian Tradition: A History of the Development of Doctrine*, vol. 4 (Chicago: University of Chicago Press, 1985).

——, *Interpreting the Bible and the Constitution* (New Haven: Yale University Press, 2004).

Penny, Andrew, "John Foxe, the *Acts and Monuments* and the Development of Prophetic Interpretation," in Loades (ed.), *John Foxe and the English Reformation*, pp. 252–77.

Pepper, Jon V., "Harriot's Earlier Work on Mathematical Navigation: Theory and Practice," in Shirley (ed.), *Thomas Harriot: Renaissance Scientist*, pp. 54–90.

——, "Harriot's Calculation of the Meridional Parts as Logarithmic Tangents," *Archive for the History of Exact Sciences*, 4 (1967–68): pp. 359–413.

Perez-Ramos, Antonio, *Francis Bacon's Idea of Science and the Maker's Knowledge Tradition* (Oxford: Clarendon Press, 1988).

Perl, Eric D., *Theophany: The Neoplatonic Philosophy of Dionysius the Areopagite* (Albany: State University of New York Press, 2007).

Pérouse, Gabriel-A., *Nouvelles françaises du XVIᵉ siècle* (Geneva: Droz, 1977).

Phillippy, Patricia, *Painting Women: Cosmetics, Canvases and Early Modern Culture* (Baltimore: Johns Hopkins University Press, 2005).

Pittock, Murray, *The Invention of Scotland: The Stuart Myth and the Scottish Identity, 1638 to the Present* (London and New York: Routledge, 1991).

Pocock, J.G.A., "Time, History, and Eschatology in the Thought of Thomas Hobbes," in *Politics, Language and Time: Essays on Political Thought and History* (New York: Athenaeum, 1971), pp. 148–201.

Pomata, Gianna and Nancy G. Siraisi (eds), *Historia: Empiricism and Erudition in Early Modern Europe* (Cambridge, MA and London: MIT Press, 2005).

Poovey, Mary, *A History of the Modern Fact: Problems of Knowledge in the Sciences of Wealth and Society* (Chicago: University of Chicago Press, 1998).

Prager, Carolyn, "'If I Be Devil': English Renaissance Responses to the Proverbial and Ecumenical Devil," *Journal of Medieval and Renaissance Studies*, 17/2 (1987): pp. 257–79.

Preston, Claire, *Thomas Browne and the Writing of Early Modern Science* (Cambridge: Cambridge University Press, 2005).

Pycior, Helena M., *Symbols, Impossible Numbers, and Geometric Entanglements* (Cambridge: Cambridge University Press, 1997).

Quillen, Carol E., *Rereading the Renaissance: Petrarch, Augustine, and the Language of Humanism* (Ann Arbor: University of Michigan, 1998).

Redondo, Augustin (ed.), *Les Discours sur le Sac de Rome de 1527* (Paris: Presses de la Sorbonne nouvelle, 1999).

Reiss, Timothy J., *Knowledge, Discovery and Imagination in Early Modern Europe* (Cambridge: Cambridge University Press, 1997).

Rhodes, Neil and Jonathan Sawday (eds), *The Computer in the Renaissance* (New York: Routledge, 2000).

Richardson, Alan and Francis F. Steen (eds), *Poetics Today*, 23/1 (2002).

Robert, Adrian, "Blending and Other Conceptual Operations in the Interpretation of Mathematical Proofs," in Jean-Pierre Koenig (ed.), *Discourse and Cognition: Bridging the Gap* (Stanford: CSLI, 1998), pp. 337–50.

Roberts, John R. (ed.), *Essential Articles for the Study of John Donne's Poetry* (Hamden: Archon, 1975).

Romanowski, Sylvie, "Descartes: From Science to Discourse," *Yale French Studies*, 49 (1974): pp. 96–109.

Ronen, Ruth, *Possible Worlds in Literary Theory* (Cambridge: Cambridge University Press, 1994).

Rorem, Paul, *Pseudo-Dionysius: A Commentary on the Texts and an Introduction to Their Influence* (Oxford: Oxford University Press, 1993).

Rosen, Edward, "Harriot's Science: The Intellectual Background," in Shirley (ed.), *Thomas Harriot: Renaissance Scientist*, pp. 1–15.

——, "Kepler's Place in the History of Science," in Beer and Beer (eds), *Kepler: Four Hundred Years*, pp. 279–85.

Rothstein, Marian, *Reading in the Renaissance:* Amadis de Gaule *and the Lessons of Memory* (Newark: University of Delaware Press, 1999).

Rowland, Ingrid D., *The Scarith of Scornello: A Tale of Renaissance Forgery* (Chicago: University of Chicago Press, 2004).

Rudnytsky, Peter L., "'The Sight of God': Donne's Poetics of Transcendence," *Texas Studies in Literature and Language*, 24/2 (1982): pp. 185–207.

Russell, Anthony, "'Thou seest mee strive for life': Magic, Virtue, and the Poetic Imagination in Donne's *Anniversaries*," *Studies in Philology*, 95/4 (1999): pp. 374–410.

Russell, Daniel S., "Perceiving, Seeing and Meaning: Emblems and Approaches to Reading in Early Modern Culture," in Peter M. Daly and John Manning (eds), *Aspects of Renaissance and Baroque Symbol Theory, 1500–1700* (New York: AMS Press, 1999), pp. 77–92.

Russell, Howard S., *Indian New England before the Mayflower* (Hanover: University Press of New England, 1980).

Ryan, Marie-Laure, "The Modal Structure of Narrative Universes," *Poetics Today*, 6 (1985): pp. 717–55.

Rybka, E., "Kepler and Copernicus," in Beer and Beer (eds), *Kepler: Four Hundred Years*, pp. 209–16.

Salmon, Vivian, *Language and Society in Early Modern England* (Amsterdam: John Benjamins, 1996).

Sandler, Florence, "The Temple of Zerubbabel: A Pattern for Reformation in Thomas Fuller's *Pisgah-Sight* and *Church-History of Britain*," *Studies in the Literary Imagination*, 10/2 (1977): pp. 29–42.

——, "Thomas Fuller's *Pisgah-Sight of Palestine* as a Comment on the Politics of Its Time," *Huntington Library Quarterly*, 41/4 (1978): pp. 317–43.

Sawday, Jonathan, *Engines of the Imagination: Renaissance Culture and the Rise of the Machine* (London: Routledge, 2007).

Schiebinger, Londa, *Plants and Empire: Bioprospecting in the Atlantic World* (Cambridge: Harvard University Press, 2004).

——, and Claudia Swann (eds), *Colonial Botany: Science, Commerce, and Politics in the Early Modern World* (Philadelphia: University of Pennsylvania Press, 2005).

Schmitt, Charles B., *John Case and Aristotelianism in Renaissance England* (Montreal: McGill-Queen's University Press, 1983).

Schofield, C.J., *Tychonic and Semi-Tychonic World Systems* (New York: Arno Press, 1981).

Schwyzer, Philip, *Archaeologies of English Renaissance Literature* (Oxford: Oxford University Press, 2007).

Segonds, A., *Introduction*, in J. Kepler, *Le Secret du Monde [Mysterium cosmographicum]* (Paris: Les Belles Lettres, 1984).

Séguin, Jean-Pierre, "Les feuilles d'information non périodiques, ou 'canards', en France," *Revue de synthèse*, 3rd ser., 78/7 (1957): pp. 391–420.

——, "La découverte de l'Italie par les soldats de Charles VIII, 1494–1495," *Gazette des Beaux-Arts*, 6th ser., 58 (1961): pp. 127–34.

——, *L'Information en France avant le périodique: 517 canards imprimés entre 1529 et 1631* (Paris: G.-P. Maisonneuve et Larose, 1964).

Seznec, Jean, *The Survival of the Pagan Gods: The Mythological Tradition and Its Place in Renaissance Humanism and Art*, trans. Barbara F. Sessions (Princeton: Princeton University of Toronto Press, 1972).

Shapin, Steven, *The Scientific Revolution* (Chicago: University of Chicago Press, 1996).

——, and Simon Schaffer, *Leviathan and the Air-pump: Hobbes, Boyle, and the Experimental Life: Including a Translation of Thomas Hobbes*, Dialogus physicus de natura aeris (Princeton: Princeton University Press, 1985).

Shapiro, Barbara, *A Culture of Fact: England, 1550–1720* (Ithaca and London: Cornell University Press, 2000).

Shea, William, "Galileo and the Church," in Lindberg and Numbers (eds), *God and Nature*, pp. 114–35.

——, "Looking at the Moon as another Earth: Terrestrial Analogies and Seventeenth-Century Telescopes," in Hallyn (ed.), *Metaphor and Analogy in the Sciences*, pp. 83–103.

Sherman, Michael A., "Political Propaganda and Renaissance Culture: French Reactions to the League of Cambrai, 1509–1510," *Sixteenth Century Journal*, 8/2 (1977): pp. 97–128.

Sherwood, Terry, *Fulfilling the Circle* (Toronto: University of Toronto Press, 1984).

Shirley, John (ed.), *Thomas Harriot: Renaissance Scientist* (Oxford: Clarendon Press, 1974).

—— (ed.), *A Source Book for the Study of Thomas Harriot* (New York: Arno Press, 1981).

——, *Thomas Harriot: A Biography* (Oxford: Oxford University Press, 1983).

Simon, Gérard, "Analogies and Metaphors in Kepler," in Hallyn (ed.), *Metaphor and Analogy in the Sciences*, pp. 71–82.

Simonin, Michel, *L'Encre et la lumière* (Geneva: Droz, 2004).

Smith, Bruce R., "Mouthpieces: Native American Voices in Thomas Harriot's *True and Brief Report* [sic] *of ... Virginia*, Gaspar Pérez de Villagrá's *Historia de la Nueva México*, and John Smith's *General History of Virginia*," *New Literary History*, 32 (2001): pp. 501–17.

Smith, Pamela H. and Paula Findlen (eds), *Merchants and Marvels: Commerce, Science and Art in Early Modern Europe* (New York: Routledge, 2002).

Smith, Robin, "Logic," in *The Cambridge Companion to Aristotle*, ed. Jonathan Barnes (New York: Cambridge University Press, 1995), pp. 27–65.

Spolsky, Ellen, "Darwin and Derrida: Cognitive Literary Theory as a Species of Post-Structuralism," *Poetics Today*, 23/1 (2002): pp. 43–62.

St. John, J.A., *Utopia: or the happy republic. To which is added, The New Atlantis, by Lord Bacon*, [...] *by J.A. St. John, Esq.* (London: Henry G. Bohn, 1845).

Stagl, Justin, *The Age of Curiosity: The Theory of Travel, 1550–1800* (Chur: Harwood Academic Press, 1995).

Stedall, Jacqueline, "Rob'd of Glories: The Posthumous Misfortunes of Thomas Harriot and His Algebra," *Archive for History of Exact Sciences*, 54 (2000): pp. 455–97.

——, *The Greate Invention of Algebra: Thomas Harriot's "Treatise on Equations"* (Oxford: Oxford University Press, 2003).

——, "Symbolism, Combinations, and Visual Imagery in the Mathematics of Thomas Harriot," *Historia Mathematica*, 34 (2007): pp. 380–401.

——, and Janet Beery (eds), *Thomas Harriot's Doctrine of Triangular Numbers: The "Magisteria magna"* (Freiburg: European Mathematical Society, 2009).

Stephenson, Bruce, *Kepler's Physical Astronomy* (New York: Springer-Verlag, 1987).

——, *The Music of the Heavens: Kepler's Harmonic Astronomy* (Princeton: Princeton University Press, 1994).

Sullivan, Garrett A., Jr., *Memory and Forgetting in English Renaissance Drama: Shakespeare, Marlowe, Webster* (Cambridge: Cambridge University Press, 2005).

Swann, Marjorie, *Curiosities and Texts: The Culture of Collecting in Early Modern England* (Philadelphia: University of Pennsylvania Press, 2001).

Sweetser, Eve, "'The suburbs of your good pleasure': Cognition, Culture and the Bases of Metaphoric Structure," in G. Bradshaw, T. Bishop, and M. Turner (eds), *The Shakespearean International Yearbook, vol. 4: Shakespeare Studies Today* (Aldershot: Ashgate, 2004), pp. 24–55.

Swerdlow, N.M., and O. Neugebauer, *Mathematical Astronomy in Copernicus's De revolutionibus* (New York: Springer-Verlag, 1984).

Tanner, R.C.H., "On the Role of Equality and Inequality in the History of Mathematics," *The British Journal for the History of Science*, 1/2 (1962): pp. 159–69.

Taylor, Charles, *A Secular Age* (Cambridge: Belknap, 2007).

Thiry, Claude, *L'Histoire immédiate: une invention du Moyen Age?* (Liège: Université de Liège, 1984).

——, "Historiographie et actualité (XIVᵉ et XVᵉ siècles)," in H.U. Gumbrecht et al. (eds), *La littérature historiographique des origines à 1500*, 1/3 (Heidelberg: Carl Winter Universitätsverlag, 1987), pp. 1025–63.

Tokson, Elliot H., *The Popular Image of the Black Man in English Drama 1550–1688* (Boston: G.K. Hall, 1982).

Tomlinson, F., *Music in Renaissance Magic* (Chicago: University of Chicago Press, 1993).

Tuscano, Pasquale, *Poetica e Poesia di Tommaso Campanella* (Milano: IPL, 1969).

Van Deusen, Neil, "The Place of Telesio in the History of Philosophy," *The Philosophical Review*, 44/5 (1935): pp. 417–34.

Vaughan, Virginia Mason, *Representing Blackness on English Stages, 1500–1800* (Cambridge: Cambridge University Press, 2005).

Vickers, Brian, "Francis Bacon and the Progress of Knowledge," *Journal of the History of Ideas*, 53/3 (1992): pp. 495–518.

Voelkel, James R., *The Composition of Kepler's* Astronomia nova (Princeton: Princeton University Press, 2001).

Vredeveld, Harry, "Deaf as Ulysses to the Siren's Song: The Story of a Forgotten Topos," *Renaissance Quarterly*, 54/3 (2001): pp. 846–82.

Wain, John, *The Living World of Shakespeare: A Playgoer's Guide* (Louisiana: Pelican, 1966).

Walker, D.P., *Spiritual and Demonic Magic from Ficino to Campanella* (Notre Dame: University of Notre Dame Press, 1975).

Washburn, Wilcomb E., *The Age of Discovery* (Washington, DC: Service Center for Teachers of History, 1966).

Wear, Sarah Klitenic, and John Dillon, *Dionysius the Areopagite and the Neoplatonist Tradition: Despoiling the Hellenes* (Aldershot: Ashgate, 2007).

Weil, Françoise, "Les gazettes manuscrites avant 1750," in Pierre Rétat (ed.), *Le Journalisme d'Ancien Régime* (Lyon, 1982), pp. 93–100.

West, William N., *Theatres and Encyclopedias in Early Modern Europe* (Cambridge: Cambridge University Press, 2002).

Westman, Robert S., "The Melanchthon Circle: Rheticus and the Wittenberg Interpretation of the Copernican Theory," *Isis*, 66 (1975): pp. 163–93.

——, "The Copernicans and the Churches," in Lindberg and Numbers (eds), *God and Nature*, pp. 76–113.

White, Hayden, *The Content of the Form: Narrative Discourse and Historical Representation* (Baltimore: Johns Hopkins University Press, 1987).

Whitehead, Alfred North, *An Introduction to Mathematics* (Oxford: Oxford University Press, 1948).

Wilken, Robert L., *The Land Called Holy: Palestine in Christian History and Thought* (New Haven: Yale University Press, 1994).

Willard, Thomas, "Donne's Anatomy Lesson: Vesalian or Paracelsian?" *John Donne Journal*, 3/1 (1984): pp. 35–48.

Williams, Arnold, *The Common Expositor: An Account of the Commentaries on Genesis, 1527–1633* (Chapel Hill: University of North Carolina Press, 1948).

Williams, George W. (ed.), *The Complete Poetry of Richard Crashaw* (New York: New York University Press, 1970).

Wind, Edgar, "The Revival of Origen," in Dorothy Miner (ed.), *Studies in Art and Literature for Belle da Costa Greene* (Princeton: Princeton University Press, 1954), pp. 412–24.

——, *Pagan Mysteries in the Renaissance* (London: Faber and Faber, 1968).

Wooden, Warren, *John Foxe* (Boston: Twayne, 1983).

Woolf, Daniel, "The Rhetoric of Martyrdom: Generic Contradiction and Narrative Strategy in John Foxe's *Actes and Monuments*," in Thomas F. Mayer and D.R. Woolf (eds), *The Rhetorics of Life-Writing in Early Modern Europe: From Cassandra Fedele to Louis XIV* (Ann Arbor: University of Michigan Press, 1995), pp. 243–82.

——, *The Social Circulation of the Past: English Historical Culture 1500–1730* (Oxford: Oxford University Press, 2003).

Zambelli, Paola, *L'Ambigua Natura della Magia* (Milano: il Saggiatore, 1991).

Index